BET 5.40

Grunch
of Giants

Books by R. Buckminster Fuller
available from St. Martin's Press

Critical Path
Tetrascroll: A Cosmic Fairy Tale
Grunch of Giants

R. BUCKMINSTER FULLER

Grunch of Giants

ST. MARTIN'S PRESS
New York

Library of Congress Cataloging in Publication Data

Fuller, R. Buckminster (Richard Buckminster), 1895–
    Grunch of giants.

    Includes index.
    1. Technology—Social aspects. 2. Civilization,
Modern—1950-   I. Title.
T14.5.F93 1983   303.4'83   82-19121
ISBN 0-312-35193-3

Design by Dennis J. Grastorf

FIRST EDITION

10 9 8 7 6 5 4 3 2 1

I dedicate this book to three women: one of the
nineteenth and two of the twentieth centuries.
First, to my great aunt, Margaret Fuller Assoli, who
with Ralph Waldo Emerson co-edited the
Transcendentalist magazine, the *Dial,* and was the
first to publish Thoreau—and herself authored
Woman in the Nineteenth Century.
I am sure Margaret would and probably does join in
my enthusiastic support and co-dedication of this
book to Marilyn Ferguson, author of *The Aquarian
Conspiracy,* and to Barbara Marx Hubbard, founder of
the Committee for the Future, for their effective
inspiration to the young world to do its own thinking
and to act in accordance therewith.

Contents

Foreword   ix

1. (Fee x fie x fo x fum)⁴   1

2. Astro-age David's Sling   7

3. Heads or Tails We Win, Inc.   18

4. Invisible Know-How, Inc.   34

5. Paper-into-Gold Alchemists   55

6. Can't Fool Cosmic Computer   77

Index   93

Sample

Foreword

There exists a realizable,
evolutionary alternative
to our being
either atom-bombed into extinction
or crowding ourselves off the planet.
The alternative is the computer-persuadable
veering of big business
from its weaponry fixation
to accommodation of all humanity
at an aerospace level of technology,
with the vastly larger,
far more enduringly profitable for all,
entirely new
World Livingry Service Industry.
It is statistically evident
that the more advanced
the living standard,
the lower the birth rate.

It is essential that anyone reading this book know at the outset that the author is apolitical. I was convinced in 1927 that humanity's most fundamental survival problems could never be solved by politics. Nineteen twenty-seven was the year when a human first flew alone across an ocean in one day.* This was the obvious beginning of the swift integration of all humanity, groups of whose members for all their previous millions of years on planet Earth had been so remotely deployed from one another that they existed as separate nations with ways of life approximately unknown to one another. It was obvious that the integration would require enormous amounts of energy. It was obvious that the fossil fuels were exhaustible. It was obvious that a minority of selfish humans would organize themselves to exploit the majority's transitional dilemmas. I was convinced that within the twentieth century, all of humanity on our planet would enter a period of total crisis. I could see that there was an alternative to politics and its ever more wasteful, warring, and inherently vain attempts to solve one-sidedly all humanity's basic economic and social problems.

That alternative was through invention, development, and reduction to the physically working stages of mass-production prototypes of each member of a complete family of intercomplementary artifacts, structurally, mechanically, chemically, metallurgically, electromagnetically, and cybernetically designed to provide so much performance per each erg of energy, pound of material, and second of time invested as to make it eminently feasible and prac-

---

*In 1944, the DC-4 started flying secret war-ferryings across both the Atlantic and Pacific oceans. In 1961, jet airliners put the ocean passenger ships out of business. In 1981, the world-around airlines flew over a billion and a half scheduled passenger miles and carried hundreds of millions of ton-miles of freight.

ticable to provide a sustainable standard of living for all humanity—more advanced, pleasing, and increasingly productive than any ever experienced or dreamed of by anyone in all history. It was clear that this advanced level could be entirely sustained by the many derivatives of our daily income of Sun energy. It was clear that it could be attained and maintained by artifacts that would emancipate humans from piped, wired, and metered exploitation of the many by the few. This family of artifacts leading to such comprehensive human success I identified as *livingry* in contradistinction to politics' *weaponry.* I called it technologically reforming the environment instead of trying politically to reform the people.*

Equally important, I set about fifty-five years ago (1927) to see what a penniless, unknown human individual with a dependent wife and newborn child might be able to do effectively on behalf of all humanity in realistically developing such an alternative program. Being human, I made all the mistakes there were to be made, but I learned to learn by realistic recognition of the constituent facts of the mistake-making and attempted to understand what the uncovered truths were trying to teach me.

In my Philadelphia archives there are approximately forty thousand articles published during the last sixty years which successively document my progressive completions of the whole intercomplementary family of scheduled artifacts. These livingry items include the following:

Tensegrity: The continuous-tension/discontinuous-compression structuring principle of Universe (i.e., stars not touching planets, electrons not touching their atomic

---

*I explain that concept in great detail in the latter part of this book. I also elucidated it in my book *Critical Path,* published in the spring of 1981 by St. Martin's Press.

nuclei) introduced to planet Earth to replace the continuously compressioned, secondarily tensioned structuring in present world-around engineering theory. Designed, 1929; prototyped, 1929.

The Dymaxion House: The autonomous, mass-producible, air-deliverable dwelling machine weighing only 3 percent of its equivolumed and equipped, conventional counterpart, single-family dwelling. Designed, 1927; modeled, 1928; helicopter-delivered, 1954.

The one-piece, 250-pound bathroom: Designed, 1928; prototyped, 1936; mass-produced in polyester fiberglass in West Germany, 1970.

Synergetics: Exploration and publishing of the four-dimensional geometrical coordinate system employed by nature. (See *Synergetics* and *Synergetics II* [New York: Macmillan, 1975, 1979].) Discovered, 1927; published, 1944.

Dymaxion World Map: Discovery and development of a new cartographic projection system by which humanity can view the map of the whole planet Earth as one-world island in one-world ocean, without any visible distortion in the relative size and shape of any of the land masses and without any breaks in the continental contours. This is the undistorted map for studying world problems and displaying in their true proportion resources and other data. Discovered, 1933; published, 1943.

World Game: A grand-strategy program developing the *design* science of solving all problems with artifacts, invented by self or others, which take advantage of all scientific and technological development through studies of their effects on the total world's social and economic affairs as ascertainable from the Dymaxion Sky-Ocean World Map. A means of assessing the feasibility of realizing various initiatives in solving world problems. Invented, 1927; applied, 1928.

Trends and Transformation Charts: These depict the total history of all the metallurgical, chemical, electro-magnetic, structural, and mechanical trendings to greater performance per given amounts of given materials, time, and energy. A compendium of all the scenarios of science and technology's evoluting advances. Chronological chart of total history of scientific discoveries and technical inventions. Chronological chart of the mining of all metals and recirculation of the scrap of those metals. Chronological charts of all major industries' performances assessed in terms of per capita human use. These charts, begun in 1928; first published, 1937, at Bureau of Standards, Washington, D.C.; published in *Nine Chains to the Moon,* 1938; published in *Fortune* magazine's tenth anniversary issue, 1940. This issue of *Fortune* went into three printings and took *Fortune* from red- to black-ink status. It changed U.S.A. and world economic health assessment from a tonnage criteria to one based on energy consumption.

The Dymaxion Omnitransport (for land, air, water-surface, and submarine travel): The first full-scale working prototype stage of which was the Dymaxion car, produced to test the crosswind ground-taxiing behaviors of an omni-streamlined, ultimately to be twin-orientable-jet-stilts-flown, wingless flying device, which would take off and land like an eagle or duck, without any prepared landing fields (similar in principle to the forty-years-later descent and takeoff, multijet system of the Apollo Moon Landing craft). Designed, 1927; prototyped, 1933.

Geodesic Domes: The unlimited-size, clear-span structures to accommodate both humanity's converging and deploying activities. Invented, 1938; prototyped, 1947. Since then, over 300,000 have been produced and installed around the world from northernmost Greenland to the exact South Pole; over 100,000 installed in children's playgrounds.

Octet Truss: The flooring or roofing structure for unlimited spanning. Designed, 1933; prototyped, 1949.

The Fog Gun: The pneumatic means of cleaning human body, dishes, clothing, etc., without plumbing's piped-in water supply. Designed, 1927; prototyped, 1949.

Compact, odorless toilet equipment: For conversion of human wastes into methane gas and fertilizer. Designed, 1928; proven in India; now being refined for production use.

Carbon blocks-inserted, copper disc-brakes: Invented and successfully demonstrated at Phelps Dodge, 1937.

Bunsen-burner-melted, water-cooled centrifuge: For processing low-grade tin ore. Invented and successfully demonstrated at Phelps Dodge, 1937.

Hanging book shelves, and other furniture: Invented, 1928; prototyped, 1930.

Modeling of all geometric developments of energetic-synergetic geometry: Including tensegrity models of all geometrical structures and the hierarchy of primitive structural systems. The minimum, all-space-filling module. The foldable, seven unique great-circle models. The tetrahelixes. Discovered, 1927; demonstrated, 1936.

Twin-hull rowing and sailing devices: Invented, 1938; prototyped, 1954.

Triangular geodesic framing of ocean-sailing hulls: Invented, 1948; successfully demonstrated, I.O.R. racing sloop *Imp,* 1979.

Very frequently I hear or read of my artifacts adjudged by critics as being "failures," because I did not get them into mass-production and "make money with them." Such money-making-as-criteria-of-success critics do not realize that money-making was never my goal. I learned very early and painfully that you have to decide at the outset whether you are trying to make money or to make sense, as they are mutually exclusive. I saw that nature has various categories of unique gestation lags between conception of something and its birth. In humans, conception to birth is nine months. In electronics, it is two years between inventive conception and industrialized production. In aeronautics, it is five years between invention and operating use. In automobiles, it is ten years between conception and mass-production. In railroading, the gestation is fifteen years. In big-city skyscraper construction, the gestational lag is twenty-five years. For instance, it was twenty-five years between the accidental falling of a steel bar into fresh cement and the practical use of steel-reinforced concrete in major buildings. Dependent on the size and situation, the period of gestation in the single-family residences varies between fifty and seventy-five years.

Because of these lags, the earlier I could introduce the conception model, the earlier its birth could take place. I assumed that the birth into everyday life of the livingry artifacts whose working conceptual prototypes I was producing would be governed by those respective-category gestation lags. I assumed my livingry inventions' progressive adoptions by society would occur only in emergencies. I called this "emergence through emergency." For all of humanity to begin to break away from its conditioned reflexes regarding living facilities (home customs and styles), allowing them to be advantaged by my livingry artifacts, would take at minimum a half century to get underway. Since this was clearly a half-to-three-quarters-of-a-century undertaking, I saw at the outset that I best not attempt it if I was not content to go along with nature's laws.

My first publication was in 1927, a bound, mimeographed book entitled *4-D,* standing for "fourth dimension." I put on the cover, "Two billion new homes will be required by humanity in the next eighty years."

Five years after I undertook the program in 1927, *Fortune* magazine in July 1932 featured my Dymaxion House in an article written by Archibald MacLeish on "The Industry Industry Missed: The Mass-Production Housing Industry." They were not aware of the complexity of the development. They had not discovered the different gestation rates of industrial evolution's diverse product categories. *Fortune* described clearly what had happened, but *Fortune*'s writers were completely unschooled in dealing with that which had not yet happened. My 1927 Dymaxion House had excited many into trying to make money by being the first in the mass-production house industry, not knowing it would take a completely new prototype and a half-century before across-the-board evolution was ready for it. A group of prominent industrialists led by one of my young fans had incorporated "General Houses," and that was all it took to get *Fortune* excited.

Thirteen years later, in 1945, *Fortune* again featured my scientific dwelling—the Fuller House prototype—only one of which I had produced, for the United States Air Force in Beech Aircraft's Wichita, Kansas, plant. *Fortune* and many others were sure that this was to be an immediate mass-production success and, more importantly to them, a "money-maker."

After Beech's production-engineering department had carefully analyzed the prototype and priced out its tooling costs, they tendered a firm bid to the Fuller House Company to produce replicas of this prototype in quantities of no less than 20,000 per year, at a base, f.o.b. Wichita factory price of $1,800, minus the kitchen and laundry equipment, to be furnished by General Electric for another $400 each. But Beech's bid required that $10 million for tooling would have to be provided by others. Beech already had Wichita banks' backing of $10 million for their post–World War II private corporate plane production.

The first Beech-produced Fuller House was widely publicized. Soon 36,000 unsolicited orders, many with checks attached, were received for the Fuller House. At this point it was discovered that no distributing industry existed. The general building contractors had none of the complex tools for several-in-one-day deliveries of the dwelling machines. None of the building codes would permit their erection. The severest blow of all was that both the national electricians and plumber's organizations said they would have to be paid to take apart all the prefabricated and pre-installed wiring and plumbing, and put it together again, else they would not connect the otherwise "ready to live in" house to the town's or city's electrical lines and water mains. They held exclusively the official license to do this by long-time politically enacted laws. No banks were willing to provide mortgages to cover the sale of the Fuller Houses.

Fortune made the mistake of assuming that "the industry industry missed" had at last come of age. But evolution's inauguration of the "livingry" industry had to wait until capitalism had graduated, from its for-centuries-held assumption that physical land property constituted capitalism itself, to the startling realization that the strictly metaphysical, technological "know-how" had become the most profitable property as the key to exploitation of the invisible industries of chemistry, metallurgy, electromagnetics, and atomics.

Reorganizing all its strategies, capitalism has now unloaded its real estate property onto the people by refusing to rent and forcing people to buy their condo or coop homes. Evolution had to wait upon the government-guaranteed, forty-year mortgage-financing of housing's costs to exceed humans' financial capability to acquire. Evolution had to wait until the U.S. mass-production of automobiles exclusively as a money-making business had been made obsolete by the technological felicity of manufacturing by other countries' producers, thus leaving U.S. productivity to reorient itself to the necessity of rehousing all humanity in mass-produced, aerospace-level-of-technology livingry.

Evolution was clearly intent on postponing the inception of the livingry service industry until humanity had graduated from its pre–twentieth century condition as a planet of remote nations to an integrated global society, all of which waited upon completion of a world-around network of highways, airlines, and telephones, and automobiles and jumbo jet airplanes. All these evolutionary events (requisite to the livingry industry) have now taken place or are about to take place in the very near future. If the political systems do not eliminate humanity with their weapons, the half-century-gestating, world-around livingry service-industry will soon be born.

•   •   •

When the leader of a chain of hand-holding skaters suddenly turns in a tight circle, he sweeps the entire chain into a circle-describing pattern, with each successive skater circling at a greater radius from him. Since the greater the radius the 3.14159 times greater the circumferential distance to be traveled, the accumulated momentum of the chain imparts ever greater acceleration to the hintermost of the original chain, which centrifugal force in turn breaks the hand-gripping and causes the hintermost individual skaters to successively break away at great speed. This fast spin-off they must cope with as best they may.

When a physics teacher wishes to demonstrate the fundamentals of wave behavior to his class, he often fixes a secure hook onto the wall and fastens the end of, for example, a ten-foot-long piece of rope to the hook. Then, going to the other end of the rope, he pulls it not quite tightly, leaving just a little slack in it. He then whips the rope end ceilingward, then floorward, followed by a final jerk ceilingward. This whips a wave into the rope. This wave goes all the way to the wall hook where it turns around and returns to the teacher's hand and stops. This demonstrates that a wave always closes back to its starting point to complete a cycle. If the teacher leaves more slack in the rope, the wave has greater height and depth lengths.

In the first million years of humans on planet Earth, we have the phase one ice-skater's spin-off pattern deploying humans to unknown, remote-from-one-another lands. In the twentieth century, we have the phase two physics teacher's wave returning upon itself to complete its cycle.

First we have the phase one, world-around deployment of humanity on foot, animals, and rafts. In phase two, we have the world-around sailing ship explorers, then the royal mail steamships, followed successively by the cable, the wireless telegraphy, the telephone, the airlines, the satellite-relayed, around-the-world wireless telephony, and the world-around skybuses: all of which, unnoticed by political society, have been inexorably, completely integrating the five-centuries-earlier, utterly remotely existing, unknown-to-one-another nations.

As a scientist I am greatly interested in all that goes on in the political-economic scene and in the impact of one unforeseen technological evolutionary event after another upon that scene. I see one minuscule computer chip completely altering the whole world scene. As the half-century of utterly unpredicted technological discoveries has occurred, accompanied by an ever greater familiarity with all the world by all humanity, I have become increasingly confident that my fifty-four-years-ago judgments were sound.

I do not look upon human beings as good or bad. I don't think of my feet as a right foot and a wrong foot. My feet are exactly the inside-out reverse of each other in physical patterning (pull the rubber glove off one hand, inside-outing it, and find it fits the other hand). Physics has found no straight lines—only waves. All courses are steered by alternate right and left veering. I see the human scene in the same way. There have to be humans to perform all the right-left evolutionary interfunctionings.

I am a student of the effectiveness of the technological evolution in its all unexpected alterations of the preoccupations of humanity and in its all unexpected alterings of human behaviors and prospects.

I do know that technologically humanity now has the opportunity, for the first time in its history, to operate our planet in such a manner as to support and accommodate all humanity at a substantially more advanced standard of living than any humans have ever experienced.

This is possible not because we have found more physical resources. We have always had enough resources. What has happened that now makes the difference is that we have vastly increased our know-how of specialized innovations, all of which invisible realization integrates to make possible success for all.

I also know that this can be realized only by a technological revolution involving total Spaceship Earth, using all the resources and know-how as an integrated regenerative system, as in the design of any successful seagoing ship or of any biological organism.

Spaceship Earth now has 150 admirals. The five admirals in the staterooms immediately above the ship's fuel tanks claim that they own the oil. The admirals with staterooms surrounding the ship's kitchen, dining rooms, and food refrigerators claim they own all the food. Those with a stateroom next to a lifeboat claim that they own the lifeboat, and so forth. They then have an onboard game called balance of trade. Very shortly the majority of admirals have a deficit balance. All the while the starboard-side admirals are secretly planning to list the boat to port far enough to drown the portside admirals, while the portside admirals are secretly trying to list the boat to starboard far enough to drown the starboard-side admirals. Nobody is paying any attention to operating the ship or steering it to some port. They run out of food and fuel. They discover that they can no longer reach a port of supply. Finis.

Humanity is now experiencing history's most difficult evolutionary transformation. We are moving away from a rooted life-style with a 95-percent rate of illiteracy. We are almost unconsciously drifting away from self-identity with our ages-long, physically-remote-from-one-another existence as 150 separate, sovereign nations. Now the uprooted humans of all nations are spontaneously deploying into their physically integrated highways and airways and satellite-relayed telephone speakways, into a big-city way-stationed, world-around living system.

We may soon be atom-bombed into extinction by the pre-emptive folly of the political puppet administrators fronting for the exclusively-for-money-making, supranational corporations' weaponry industry of the now hopelessly bankrupt greatest-weapons-manufacturing nation (the U.S.A.).

If not bomb-terminated, we are on our ever swifter way to becoming an omni-integrated, majorly literate, unified Spaceship Earth society.

The new human *networks*' emergence represents the natural evolutionary expansion into the just completed, thirty-years-in-its-building, world-embracing, physical communications network. The new reorienting of human networking constitutes the heart-and-mind-pumped flow of life and intellect into the world arteries.

•　　•　　•

The world-integrating networking self-multiplies and accelerates. Never traveling as a tourist, I myself have been induced into forty-eight complete encirclements of our planet and everywhere I go I meet more and more people whom I have met elsewhere around the world. Ever more widely traveling, literate, well-informed individuals discover that they, and an ever faster increasing number of other humans, are becoming intuitively aware that life is breaking them out of the ages-long, anonymous life-patterning of the beehive drone. They experience newborn hope that humans have indeed a destiny of individual significance complementary to the integrity of other individuals.

The networking accelerates as does light in Einstein's equation $E = Mc^2$. The lower-case c is the symbol for the linear speed of light, 186,000 miles per second. When not reflectively focused, light expands omni-radially as a sphere. The rate of surface growth of a spherical wave system is always the second power of the linear (radial) growth speed. That is why it is c^2 in Einstein's equation, which means 186,000 × 186,000 miles of spherical surface growth per second. Since human thought can calculate in minutes what it takes light to travel in one year, it may be that thought itself expands outwardly in all directions at a speed even faster than light—maybe in no time at all—to inter-network the people of our eight-thousand-mile diameter spherical space home.

As the networking accelerates humanity into a spherically embracing, spontaneous union, yesterday's locally autonomous, self-preoccupied governments are left in the exclusive control of yesterday's most selfishly successful and entrenched minorities. The present U.S.A. 1982 administration was elected by the votes of only one-seventh of the U.S. population and it spent $170 million—more than five times the money raised by their opponents—to buy their victory. The networks' people, aware that a U.S. presidency costs $50 million, a senatorship $10 million, and a representative's seat $5 million, observe that the TV era governments are corrupt, wherefore they spontaneously abhor and abstain from further voting.

Gradually discovering that the networking abandonment of the voting booth was the true cause of their claimed "overwhelming majority," the incumbent administration, fearful of a potential rejective voting tidal wave of the inter-networked world people, will probably try in vain to block networking. Because networking is apolitical and amorphous, it has no "cells" to be attacked, as did the communism of former decades. The fearful sovereign nation politicos will find that trying to arrest networking is like trying to arrest the waves of the ocean.

As elucidated in *Critical Path,* the net resultant developmental patterning of human affairs is always the progressively integrating product of:

(1) cosmic evolution;
(2) the integrated evolution of myriads of individual human fear-and-longing-motivated initiatives.

Both *number one evolution* and *number two evolution* conform strictly to cosmic laws such as those governing gravity, radiation, and DNA-RNA's biological, species and individual, formation-and-growth design governance.

Number one evolution directly follows the cosmic laws.

The individuals of *number two evolution* are generally unaware of the cosmic laws and bumble their myriads of individually different ways into progressively more complex scenarios, all of which are ultimately and all-unknowingly both intercomplementary and governed by the same cosmic laws controlling *evolution number one's* two interweaving components, i.e. its *certainties* and its trial-and-error-conducted *experimental initiatives.*

Humanity is an *experimental initiative* of Universe. The experiment is to discover whether the complex of cosmic laws can maintain the integrity of eternal regeneration while allowing the mind of the species *homo sapiens* on the little planet Earth to discover and use some of the mathematical laws governing the design of Universe, whereby those humans can by trial and error develop subjectively from initial ignorance into satisfactorily informed, successful local-Universe monitors of all physically and metaphysically critical information and thereby serve objectively as satisfactory local-Universe problem-solvers in sustaining the integrity of eternally regenerative Universe.

Trial-and-error-evolved steering wisdom does not accrue to shifting the rudder angle violently leftward from an equally violent rightward-course-steering error. Evolutionary advance in trial-and-error steering systems is accomplished by successively more delicate leftward-rightward correcting of swing-over error. This progressive reduction of mechanical momentum must however be preceded by forthright acknowledgment of the error and thereby of the truth which was being hidden by the erroneous assumptions of yesterday's false self-pride rationalizing.

The function of local-in-Universe critical information-gathering and local-in-Universe problem-solving is manifest in the forward cockpit of all great airplanes. When the door to the pilots' and engineers' compartment is left open, you may see a myriad of instruments covering all the walls and ceilings of the cockpit.

The dials of the cockpit instruments report optically

and accurately all the stress, strain, heat, pressure, veloc-
ity, ratios, and other significant conditions of all critical
parts of the airplane's air-frame, fuselage, wings, rudders,
landing gear, power plant, baggage compartment, and
passenger space, as well as of the interior and exterior
temperatures, atmospheric pressure, etc., as well as all the
instrumentally reported information regarding altitude
and the ever-changing geography of clouds, wind forces,
and directions, as well as all the electromagnetically received information regarding directions of all surrounding
airports and the flight path beam bearings, etc.

Through comprehensively synchronized control of this
instrument-reported information the captain may put the
ship on "all automatic" flight as his assistant pilots and
engineers watch those dials and make appropriate automatic control adjustments in response to the changing
information. It is by virtue of this information that the
pilots and engineers are able to serve as local problem-
solvers in support of the integrity of their passengers' and
their own regenerative living integrity.

This same critical information-gathering and local
problem-solving in sustainment of the integrity of regenerative systems is also performed by the electromagnetic
instrumented airport tower controllers, who deal with a
myriad of variables of wind and speed directions, and
aircraft holding, approaching, landing, taking off, holding-
ready-to-take-off, on-field-runways-taxiing planes, each
plane worth millions of dollars, each packed full of beyond
priceability human cargoes.

All these high-velocity, vast risk, complex pattern controls of pilots, controllers, and engineers are special-case
patterns of the generalized local information-gathering
and local problem-solving in support of the integrity of
eternal regeneration.

All of the ecology on planet Earth, all the biosphere of

planet Earth as well as the radiated entropy and the gravitationally coordinated-and-coherred syntropy of the solar system and the Milky Way galaxy and the two billion other known galaxies are all special-case manifests of the local-in-Universe information-gathering and problem-solving in support of the synergetic integrity of overlappingly episoded, eternal scenario Universe.

The function of local-Universe information-gathering and local-Universe problem-solving is a generalized problem-and-solution complex, the solution of which is mathematically expressed as the updated Einstein equation $E = 2Mc^2$.

Recognizing that:

(1) this information-gathering and problem-solving function of humans on little, local planet Earth is a cosmically generalized function;

(2) our experiencing of the coordinate integrity of nature's pattern of initiating the growth of myriads of ecologically intercomplementary biological species is scientifically sorted out; and

(3) nature's method of trial-and-error evolvement of the successive biological types prospering most successfully under each unique evolutionarily progressive environmental change demonstrates that preferred-environment technology eliminates survival only of the fittest;

we come to the full realization that the failures of humans on planet Earth to fulfill satisfactorily and faithfully their generalized information-gathering and local problem-solving in support of eternal regeneration of Universe simply means death of this particular planetary installation of mind-endowed individuals. The failure of humans means the function must be performed in local-Universe by other phenomena capable of reliably serving the information-agglomerating and problem-solving function. The eternal Universe show must go on.

When individuals shunt the comprehensive cosmic regeneration into exclusive advantaging of only their own survival and enjoyment and succeed in prolonged local short-circuiting of cosmic regenerativity, they disqualify the invention "human" as a reliable function of regenerative Universe. They are just as irresponsible in the cosmic system as the company employees who pocket the cash register contents for their own account. This is cosmically true of a childless multimillionaire maneuvering himself into a position to make a big profit involving "hard-headed," absolutely selfish decisions that will knowingly and legally deprive many others of survival necessities—"to hell with the next generation"—which deal will win him the applause of other powerfully rich individuals because it makes them feel more comfortable about their own summa cum selfishness.

If you ignorantly believe there's not enough life support available on planet Earth for all humanity, then survival only of the fittest seems self-flatteringly to warrant magna-selfishness. However, it is due only to humans' born state of ignorance and the 99.99 percent invisibility of technological capabilities that they do not recognize the vast abundance of resources available to support all humanity at an omni-high standard of living.

We have now scientifically and incontrovertibly found that there is ample to support all humanity. But humanity and its leaders have not yet learned so in sufficiently convincing degree to reorient world affairs in such a manner as to realize a sustainable high standard of living for all.

There are three powerful obstacles to humanity's realization of its omni-physical success:

1. The technical means of its accomplishment exist altogether in the invisible realms of technology.
2. The experts are all too narrowly specialized in developing the invisible advance to envision the synergetic significance of integrating their own field's advances with other fields' invisible advances.
3. The utterly different, successful ways of metabolic accounting, dwelling, self-employing, cooperating, and enjoying life are unfamiliar and nonobvious.

Because of ancient arms-accomplished seizure of land by the most physically powerful and the subsequent arms-induced blessing of the seizures by power-ordained "ministers of God," royal deeds to land were written as assumedly God-approved and -guaranteed covenants.

Landlordism, first woven into the fabric of everyday life by royal fiat and thousands of years of legal process precedent, has become an accepted cosmic phenomenon as seemingly inevitable as the weather. Humans have learned to play many of its games.

Land "ownership" and its omni-dependent comprehensive thing-ownership involvements and their legal-documents-perpetuations constitute the largest socioeconomic custom error presently being maintained by a large world affairs–affecting segment of humanity.

Nothing new about all that. But what is new is that humanity has gone as far as it can go with this significant error and is in final examination as to whether it can free itself from its misconditioned reflex straightjacket in time to pull out of its greatest-in-all-history, error-occasioned tailspinning into eternity.

We do have both the knowledge and the technical means to do so if we do it quickly enough. That is what this book is about.

Grunch
of Giants

Chapter 1

[Fee X fie X fo X fum]^4

Fee-fie-fo-fum
I smell the blood of a Britishman
Be he alive or be he dead
I'll grind his bones
To make my bread.

There is no dictionary word for an army of invisible giants, one thousand miles tall, with their arms interlinked, girding the planet Earth. Since there exists just such an invisible, abstract, legal-contrivance army of giants, we have invented the word *GRUNCH* as the group designation—"a grunch of giants." GR-UN-C-H, which stands for annual GROSS UNIVERSE CASH HEIST, pays annual dividends of over one trillion U.S. dollars.

GRUNCH is engaged in the only-by-instruments-reached-and-operated, entirely invisible chemical, metallurgical, electronic, and cybernetic realms of reality. GRUNCH's giants average thirty-four years of age, most having grown out of what Eisenhower called the post–World War II "military-industrial complex." They are not the same as the pre–World War II international copper or tin cartels. The grunch of giants consists of the corporately interlocked owners of a vast invisible empire,

1

which includes airwaves and satellites; plus a vast visible empire, which includes all the only eighteen-year-old and younger skyscraper cluster cities around the world, as well as the factories and research laboratories remotely ringing the old cities and all the Oriental industrial deployment, such as in Taiwan, South Korea, Malaysia, Hong Kong, and Singapore. It controls the financial credit system of the noncommunist world together with all the financial means of initiating any world-magnitude mass-production and -distribution ventures. By making pregraduation employment contracts with almost all promising university science students, it monopolizes all the special theoretical know-how to exploit its vast inventory of already acquired invisible know-how technology.

Who runs GRUNCH? Nobody knows. It controls all the world's banks. Even the muted Swiss banks. It does what its lawyers tell it to. It maintains technical legality, and is prepared to prove it. Its law firm is named Machiavelli, Machiavelli, Atoms & Oil. Some think the second *Mach* is a cover for *Mafia*.

GRUNCH didn't invent Universe. It didn't invent anything. It monopolizes know-where and know-how but is devoid of know-why. It is preoccupied with absolute selfishness and its guaranteed gratifications. It is as blind as its Swiss banks are mute. Much, much more about GRUNCH later on.

•   •   •

When blimp photographs are taken of giant stadia packed full of rock-concert or football fans, we get an idea of what 100,000 people look like. We all think of Hiroshima as the worst single killing of humans by humans. That was about a 75,000-capacity-coliseum-full. Each day of each year, year after year, a 75,000-capacity-stadium-full of around-the-world humans perish from starvation or its side effects, despite an annual average 5-percent world food-production overage of the amount of food adequate for the total world's population. This daily kill of innocents dwarfs the awful Auschwitz killing.

GRUNCH did not bring this about, but it could very profitably bring it to an end. Just because it is possible does not mean that it is easy. With the computers' guidance, however, and some executive vision, courage, initiative, and follow-through, it can be done very profitably in terms of money and lasting kudos for GRUNCH and prohumanity enterprise. It would cost only 3 percent of Grunch's annual dividend earnings to not only feed all those now starving to death but also to alleviate the dire poverty around the entire planet, since the population explosion is occurring strictly amongst impoverished people. Such a world initiative on the part of Grunch would eliminate one of the two great threats to humanity's continuance on planet Earth: nuclear bombing and overpopulation.

The great communism vs. capitalism, politico-economic world stand-off assumes a fundamental inadequacy of life support to exist on our planet. So too do the four major religions assume that it must be you or us, never enough for both. Jointly the two political camps have spent $6.5 trillion in the last thirty-three years to buy the capability to kill all humanity in one hour.

Jointly, we Earthians have always had adequate physical resources to take care of all humanity but lacked the metaphysical know-how resources with which to employ effectively the Earth's physical wealth. Adequate know-how could only accrue through trial-and-error experience combined with synergetically acquired wisdom, altogether employed with absolute faith in the intellectual integrity omni-lovingly governing regenerative Universe. However, in 1970 our cornucopia of ever more swiftly accruing

know-how overflowed and its content integrated synergetically, so that we may now care for each Earthian
individual at a sustainable billionaire's level of affluence
while living exclusively on less than 1 percent of our
planet's daily energy income from our cosmically designed
nuclear reactor, the Sun, optimally located 92 million safe
miles away from us and safely interlinked with us by
photosynthesis, wind, rain, wave, and all other weather
behaviors.

In technology's "invisible" world, inventors continually
increase the quantity and quality of performed work per
each volume or pound of material, erg of energy, and unit
of worker and "overhead" time invested in each given
increment of attained functional performance. This complex process we call progressive ephemeralization. In
1970, the sum total of increases in overall technological
know-how and their comprehensive integration took humanity across the epochal but invisible threshold into a
state of technically realizable and economically feasible
universal success for all humanity.

This actual but invisible threshold crossing began in
1969 when humans' scientific knowledge and technological ingenuity, backed exclusively by adequate citizens'
tax-raised government financing, learned how to do so
much with so little as to be able to place humans on the
Moon and return them safely to Earth. Other typical 1970
to 1980 manifests of our option to do so much with so little
as to be able to take care of all humanity were:

1. The single-flight delivery and installation of a 140-
    foot-diameter, 23,000-square-foot-floor-space, stain-
    less steel and aluminum geodesic dome at the math-
    ematically exact South Pole of our planet, together
    with its capability of carrying the snow loads of com-
    plete burial;

2. The rocket-launched satellites able to relay Earth-around TV and other programs;
3. The solar system's planetary inspection by TV-communicating, Earth-dispatched explorer satellites;
4. The computer revolution, and its progressive miniaturization;
5. The laser-beam and its many capabilities, such as its color-TV-reading of polished disc records;
6. MacCready's successful human-muscle-powered, over-the-English-Channel flight; and
7. His subsequent Paris-to-England, exclusively by direct-Sun-powered flight; and finally,
8. That MacCready's ninety-five-foot-wingspan plane weighed only forty-five pounds due to its carbon-fiber-alloy structuring and mylar skinning.

In 1970 it could, for the first time, be engineeringly demonstrated that, applying the most advanced know-how to the conservation and use of the world's resources, we can, within ten years of from-killingry-to-livingry reoriented world production, have all humanity enjoying a sustainably higher standard of living than any humans have ever heretofore experienced. It could further be demonstrated that we can do this while simultaneously phasing out all further Earthians' use of fossil fuels and atomic energy.

Humanity is so specialized and these epochally significant technological facts are so invisible that it seems an almost hopeless matter to adequately inform humanity that from now on, for the first time in history, it does not have to be "you *or* me"—there is now enough for "both" —and to convince humanity of this fact in time to permit it to exercise its option and save itself.

There is now plenty for all. War is obsolete. It is imperative that we get the word to all humanity—RUSH—before someone ignorantly pushes the button that provokes pushing of all the buttons.

What makes so difficult the task of informing humanity of its newborn option to realize success for all is the fact that all major religions and politics thrive only on the for-all-ages-held, ignorantly adopted premise of the existence of an eternal inadequacy of life-support inherent in the design of our planet Earth.

That it is possible for us all to win—and how—is what *Grunch of Giants* is about.*

---

*Grunch of Giants is an intimately related sequel to *Critical Path,* published by St. Martin's Press, New York, 1981.

Chapter 2

Astro-age David's Sling

In each herd of wild horses there is a king stallion. Every once in a while a young stallion is born bigger and more powerful than the herd's other colts. When the new big colt matures, the king stallion engages him in battle. Whichever one wins becomes the inseminator of the herd's mares. Darwin saw this phenomenon as the way in which nature contrives to maintain the strongest, best coordinated, most alert, and fastest strains in the species. Twentieth-century racing stable operators progressively inbreed the fast-running genes. Nature uses this progressive, only-by-generation-to-generation, (DNA-RNA)-genes-concentrating method in selecting, evolving, and maintaining the physically fittest biological types to serve the vast variety of planet Earth's omni-intercomplementary ecological regeneration functions to become operative under each and every uniquely variant set of environmental conditions.

Nature employs the same solely-by-survival-through-successive-generations, genes-concentrating principle when introducing humans into the complex ecological scheme of intercomplementary regenerativity of life on planet Earth.

While nature undoubtedly initiated the installation of humans on planet Earth with semi-giant leaderships, she had eventually to disclose to humans through direct experience lessons that human muscle is naught as compared to the competence of the mind-directed brain.

• • •

I acquired one of the most important
Of my life's working assumptions
When I undertook to answer
My own 1927 self-questioning:
"Why have humans been included
In the design of Universe?"
My hypothetical answer of 1927
Was, and as yet is:

What impresses me most
Is the experientially demonstrable fact
That all living organisms
Other than humans
Have some organically integral equipment
That gives them some inherent
Physical advantage
In coping with special environmental conditions—

A plant that can and does thrive
Only under dense Amazon
River jungle conditions—
A bird that can fly beautifully

While in the sky
But which cannot divest itself
Of those wings
While awkwardly walking—
The fish having equipment
To extract oxygen from the water
That dies out of water.
Common with many creatures
Humans have brains.
Brains of humans and other creatures
Are always and only
Coordinating and memory-storing the information
Reported to their brains
By internal and external
Sensing devices
Regarding each special-case systemic experience.
But humans are given mind's access
To objectively realizable mathematical principles
whereby
Humans can produce their own wings to
Outfly all the special-case, integrally winged birds.

In addition to their brains
Humans have minds,
Possessed by no other
Known organism.
Weightless, nonphysical minds
Are concerned with discovering
The interrelated significance
Of all-time humanity's
Thus-far-experientially-discovered
And experimentally-verified inventory
Of ever-experientially-redemonstrable,
Only-mathematically-expressible
Cosmic design laws

And of those laws' governance
Of the multi-alternative freedoms of realization
As mathematically incisive,
Omnirational, variously magnituded,
Structurally associative and disassociative,
Nonsimultaneous,
And only omnicomplexedly intercomplementary,
Always and only overlapping episodes
Altogether essential
To eternally regenerative
Scenario Universe.

HUMANS:
IF
Successfully evolved
Physiologically, psychologically, and philosophically
From their born-naked,
Helpless-for-months,
No-experience state
Of absolute ignorance
To be progressively educated
As driven only by innate
Hunger, thirst, procreative instinct, and curiosity
Into initiative-takings
Can thereby discover frequent errors
Of assumption, identification, or execution
Wherewith, if the individual's innate courage
And sense of the importance of truth
Are greater than the sense of pride
Of the individual,
Error is admitted
Thus only inadvertently uncovering
That which is true
Which discovered truth may prove to be
Both physically and metaphysically

Inspiringly advantageous information
Suggesting ways of progressively improving
Physical life-support systems
And their environmental realizations
Together with their operational information agenda
Whereafter, encouraged by experiencing
The ensuing, more favorable environmental
circumstances
Accruing only to their ever abiding by the spontaneous
Self-admonitions springing intuitively
From their innate love of the truth,
The thus-inspired individuals
Persevere with integrity
Throughout some hundred thousand generations
Of such only-by-trial-and-error conditioning
With each generation's intermatings
Of those mutually surviving
Under the evolving environment
And mutually educated exclusively

By such artifacts-accomplished,
Creative conversions
Of negative into positive circumstances
And through those regenerative matings
To concentrate the DNA-RNA,
Exclusively angle- and frequency-controlled,
Structural and mechanical design programmings
Of the creatively imagining faculties
And their corresponding crafting dexterities
Which with their inventing of words
Their metaphysical tools
And electromagnetic-spectrum
Communications accomplishments
Might in time
Attain and sustain

A semi-divine level
Of exclusively artifacts-realized,
Creative design wisdom
Adequate to render the cosmic environment
Healthily supportive of all humanity

THEN
Shall humans discover
That they have been
Included in Universe
To function:
First, as local Universe information-harvesters;
Secondly, as critical information-winnowers;
Thirdly, as generalized
Patterns-and-principles discoverers;
And fourthly, employing those principles objectively
To serve as local Universe problem-solvers
In support of the integrity
Of eternally regenerative,
Only overlappingly inter-episoded
Scenario Universe.

●   ●   ●

In order that humans might so evolvingly function, they were first given brains and then access to mind.

As already mentioned but worth repeating, human brains, as with those of many creatures, function always and only as coordinators of all the sense-apprehended information regarding each special-case temporal experience, all of which special-case experiences have beginnings and endings.

In contradistinction to brains, which are constructed of physical matter, the weightless, matterless, metaphysical mind has the unique capability, from time to time, to discover eternal interrelationships existing invisibly be-

ASTRO-AGE DAVID'S SLING / 13

tween special-case experiences, which interrelationships cannot be discovered by any or all of the brain's physical sense systems—for instance, the mathematical law governing gravity's invisibly cohering not only the Sun and its multi-millions-of-miles-apart planets as discovered by the weightless invisible conceptual thought-relaying from the mind of Kepler to Galileo to Newton and as also cohering the never-anywhere-intertouching parts of local Universe systems of galaxies, and electrons remote from their nuclei.

The relative interattractiveness invisibly operative between any two remote-from-one-another cosmic bodies, as compared to any other pair of cosmic bodies, equally distanced from one another, is proportional to the multiplicative product of the respective couple's masses, and the interattractiveness of any pair of celestial bodies varies inversely as the second power of the distance between them. Halve the intervening distance and the interattractiveness increases fourfold.

Human minds were given the semi-divine capability to discover and employ some of the only-mathematically-communicable eternal laws governing the design of eternally regenerative Universe itself.

When I was born in 1895 humanity was 95-percent illiterate and needed leaders. Today the situation is reversed. Humanity is now 65-percent literate and capable of doing its own thinking, decision-making, and initiative-taking.

•   •   •

Returning to the genetic evolution and the history of humans aboard our planet, we observe that only by the genes-concentrating of successive generations of survivor-matings were design-produced the best all-around average-size humans for average as well as special environmental conditions. Both giants and pygmies were design-evolved for coping with extremes of environment.

Bare-handed giants could physically overwhelm both bare-handed average humans and pygmies, except when the littler ones escaped into caves or thickets through entryways too small for giants.

People think of the size of people in terms of height. Mathematics shows, however, that if we double human height while maintaining equal proportionment we have four times as much surface (skin) and eight times as much volume (flesh and bone—weight). "Twice as big" is really eight times as big. Giants were indeed overpowering.

Tools are the only-intermittently-used, noncorporeal extensions of integral functioning capabilities of biological species. Spider webs are tool extensions of spiders. Nests are short-term tool extensions of birds whose regenerative functioning occupies only a minor fraction of their lifetime activity. In order to be light enough to be able to fly, the birds must physically separate out all those of their overall essential lifetime functions not continuously required in their survival and development. Momentarily containing both embryo and nutriment, eggs are externalizable, system-separable, new-life-gestation tools which together with insulating nests provide the means by which the mother bird may fly unencumbered to seek out the worm- or insect-packaged energy-intake-as-heat to be transferred to the embryo inside the heat conductive eggshell in its heat-conserving nest and do so before the eggs become too cold. Later she brings the worms and insects directly to the hatched chicks themselves, secure in their heat-conserving nest—an environment-controlling tool.

Nests and eggs are indeed tools, as is the womb—an only-once-in-a-while, carried-with-mammalian, new-life-production tool. Mammals don't have to fly so they can carry tools integrally, internally, as do they also carry their hearts, livers, and other continuously interlinked high-frequency-of-use tools. All tools exist in Universe only as essential functional components of development programs of living organisms.

A cut-off finger can be swiftly stuck back onto the hand to seemingly function again as an apparently integral part of the organism. The comprehensive fact, however, is that nothing in Universe touches anything else. There are no *solids*. There are, in fact, no *things*. There are only complex critical-proximity and -frequency, unique *event* aggregates interoperative in pure principle. The event *electron* is as remote from its nuclear events as is the Moon from the Earth as size-referenced to their respective event diameters.

Biological organisms, like all systems in Universe, are constituted of locally interregenerative functions in pure principle. So too with all species. Humans have a vast range of overall essential functionings whose frequency and duration of use can be developed either as integral or non-integral, inorganic or organic tool-extensions of their pounding, cutting, scratching, marking, formative, or transformative and transportive functions of survival and development.

Simple tool-inventing, such as picking up a stone to throw in self-defense, requires only the brain's instinctive functioning. Almost everywhere stones lie ready to serve as heavier "fists" for powerful, but only occasionally needed, punching, pounding, or smashing. Mind, which alone comprehends the complex interaction of principles, is required to anticipatorily invent stone-throwing slings, spring-loaded catapults, or bow-and-arrows archery.

Tools of self-defense or aggressive warfare, called weapons, frequently embody principles that can be constructively rather than destructively employed. Beyond even that, they can solve positively the originally negative, aggression-aggravating problem—a basic precept of design science. Conversely, ballpoint pens, carried with us in our breast pockets because so frequently used, can be employed either constructively to write a life-saving prescription or destructively as a dagger. My hearing aids and my eyeglasses are more permanent components of my awake hours than are millions of daily dying-and-being-replaced protoplasmic cells of which my flesh is constructed—otherwise I could not lose a pound a day. My eyeglasses and hearing aids together weigh only a quarter of a pound. The same tool can become the extensions of many individuals, and for longer years than human generations. Bridges and transcontinental highways are universally social tools.

Airplanes are mutually interchangeable human flying tools. There are metaphysical as well as physical tools. Mathematics are metaphysical tools. Inter-relevant metaphysical informations are tools. The Massachusetts Institute of Technology's Department of Mathematics states officially: "Mathematics is the science of structure and pattern in general." Physical structures are tools. Mathematics makes possible the synergetic organization of metaphysical information which can be progressively objectivized as pattern-controlling or evolving tools systems which can advantageously alter the physical environmental circumstances of human existence.

The comprehensive name for the omni-interrelated significance of all physical and metaphysical tools is "technology." All technology is governed by generalized physical laws. All of these can be physically demonstrated and mathematically expressed. The physical Universe itself is omnitechnology—a complex of various frequencies of intercomplementary functions altogether producing nonsimultaneous, multi-frequenced, multimagnituded, only overlappingly inter-episoded, eternal regeneration.

Man-made laws, legal agreements, royal fiats are not

tools. Man-made laws and customs are not technology. They are political power ploys, originally instituted only by-physical-might-asserted and -sustained "rights." Corporations are not tools. They are conventionally accepted power-proclaimed legal contrivances.

There are two fundamental types of tools: those which can be produced by one human unaided in any way physically or metaphysically by any other humans and those which could not be invented, developed, or operated by one human alone. The latter, multi-human-involvement tools, are the industrial tools.

The solo-evolved tools are craft tools. The solo-developed craft tools can be complex and even powered, as by a harnessed creature or a dammed stream of water. Craft tools used as weapons make it possible for physically small humans to overwhelm physically either big humans or even bigger animals—tiger pitfall traps are one example.

This brings us to the slaying of the giant Goliath by David and to the generalized principle of brain-mastered brawn by mind-mastered brain, of the metaphysical mastering the physical.

This in turn brings us to the present confrontation of humanity by the Grunch of Giants—the supranational corporate conglomerates—the greatest giants in all history invisibly "Rough-Riding" planet Earth. While you can see their skyscrapers and factories, these are only the physical properties occupied by the human-drone workers employed by the elusively invisible corporate conglomerates.

Chapter 3

Heads or Tails We Win, Inc.

Corporations are neither physical nor metaphysical phenomena. They are socioeconomic ploys—legally enacted game-playing—agreed upon only between over-whelmingly powerful socioeconomic individuals and by them imposed upon human society and its all unwitting members. How can little humans successfully cope with this greatest of all history's invisible Grunch of nonhuman Giants? First of all, we humans must comprehend the giants' games and game-playing equipment, rules and scoring systems. But before we can comprehend their game-playing, we must study the history and development of giants themselves.

One of my many-years-ago friends, long since deceased, was a giant, a member of the Morgan family. He said to me: "Bucky, I am very fond of you, so I am sorry to have to tell you that you will never be a success. You go around explaining in simple terms that which people have not been comprehending, when the first law of success is, 'Never make things simple when you can make them complicated.' "

So, despite his well-meaning advice, here I go explaining giants.

•   •   •

In addition to the B.C. David and Goliath theme, we have the A.D. 800 story of Roland (Childe Roland), legendary son of Charlemagne's sister Gilles. There are many poetical chronicles of young Roland's *enfances* (a very young person's heroic exploits), such as vanquishing giants—one named Ferragus and another Eaumont. From the eighth to the seventeenth century, many variations of the story occur, published in Latin, Italian, French, and English.

Much esteemed in Italy, Roland was known there as "Orlando Furioso"—the order of the name's first two letters is reversed from *ro* to *or*—as immortalized in the A.D. 1502 poem by Ludovico Ariosto.

The first comprehensive chronicling of Roland was written in Latin by Turpin, Archbishop of Reims, before A.D. 800. Roland (or Orlando) is mentioned by Dante in his *Paradiso* and is the subject of songs sung at the Battle of Hastings in the *Chanson de Roland* (c. A.D. 1100). Shakespeare mentions him in *King Lear.*

With the advent of radio and television, the children's Mother Goose–type storybooks of yesterday have been progressively abandoned. Few people today are familiar with the thousand-year-old story of the roaring of the giant as Roland approached his tower: "Fee-fie-fo-fum/ I smell the blood of an Englishman/ Be he alive/ Or be he dead/ I'll grind his bones/ To make my bread."

Supreme horse-mounted monarchs in the days of Roland could and did award vast hunting and farming lands to their horse-mounted blood kin and military henchmen, who together hunted their lands and had them cultivated by on-foot, tithe-paying tenant farmers.

In ancient North China a new kind of giant had developed long, long before Roland's time—a three-component-parts giant, i.e., the little *man,* with a *club,* mounted on a *horse*—who could and did overwhelm the big, on-foot, tribe-leading shepherd. This new composite giant, the horse-mounted bully, could divert to his sole advantage as much as he wanted of the life-support productivity of the on-foot peasantry.* The horse-mounted, club-wielding bully asserted—as do the twentieth-century racketeers—that he owned the land on which the shepherds were grazing their sheep or the farmers were growing their crops. There was no way in which the shepherd could realistically contradict the bully. Each night, many of the shepherd's sheep disappeared until the shepherd agreed to "accept" the horse-mounted bully's "protection." This was the origin of "property." The most powerful amongst the leaders of gangs of horsemen became the emperor.

The emperor rewarded his henchmen with *deeds* to the land in proportion to the deeds at arms they performed for him.

There is no historical record of religion founders who have been so bold as to assert that God had deeded land to anyone. History shows that religious leaders have, however, frequently complied with their king's instructions to plant a cross or other symbol of God's approval of their king's sword-accomplished vast lands-seizure and ownership-claiming.

Pa ys = land; ped = foot = ped ant = pa ys antry = peasantry = combination of *on the land* and *on foot* = *pa* y of lands = *pa* of patriot = *pa* of pagans = *pa* tois = *pa-*gan, *pa* gan peasantry.

Over thirty thousand years ago, these prehistoric horse-mounted "landowners" began expanding their territory northwardly and westwardly beyond the Himalayas into Mongolia and then ever westward into Europe.

Also, starting at least thirty thousand years ago, South Pacific islanders and south and northeast continental Asians came to the West Coast of North, Central, and South America from the Orient by rafts swept along by the Japan Current. Many if not most of the rafted Southeast Asians colonized the West Coast of the Americas and islands of the east Pacific. The current then returned some of the rafters to Southeast Asia, as Thor Heyerdahl demonstrated with his raft *Kon-Tiki.* This circum-Pacific ovaling of the Japan Current raft-travel outlined the Polynesian world within which was spoken a commonly based language. The Polynesians became the world's water people. Polynesia comprised more than one-quarter of the planet Earth. The great West Coast mountain ranges and deserts slowed both the North and South American coastal, raft-landed colonists' eastward migrations. Landed at many North and South American coastal points from Alaska to Chile, these raft-landed Polynesians separated into many groups as they moved eastward over many routes to both North and South America, to become known as the American Indians.

As water people, the Indians assumed that the "Great Spirit" (not an anthropomorphic God) gave them fishing, hunting, and cultivating rights, but never ownership of land. Obviously, to them, only the Great Spirit could own the land. Centuries later the Indians thought they were selling the Europeans only fishing and hunting licenses, not property rights. These were water people. No sailor can think realistically of "owning" a specific area of the ever transforming oceanic waters. Many pirates tried vainly to do so.

• • •

We have, historically, two prime, oppositely directed world-encirclings, both starting about thirty thousand years ago: (1) from the Orient via water, eastbound from Southeast Asia, and (2) westbound via land from northeast Asia. Mastery of all the sea finally went to one land-based nation after another.

• • •

Millennia after the first club-swinging Oriental horseman claimed land ownership, the man on the horse westbound from the Orient to Europe became helmeted and armored in metal. Due to the horses' weight-carrying limit and the penalty of weight on the horses' speed, the most effective of the horse and armored riders was, like the present-day jockey, the wiry, strong, little man. Inspection of the European museums' armor discloses the diminutive size of the most successful knights. The main significance of what we are learning is that, to the man on foot, the horse-mounted and armed men became a new and formidable "giant."

Because the armored knight required many helping hands to mount him and maintain his horses and arms, he had to have their goodwill and support lest his helpers overwhelm him when dismounted and encased in his armor. As a consequence, the little, wiry man in horse-mounted armor frequently became the champion of traveling bands of the little people. The little armored knight was more maneuverably effective than the armored giant when the latter's multifolded weight overburdened his mount.

As a special consequence of this trending, we have the nongigantic, successfully armored King Arthur's Round Table Knights, who used their mounted and armed might to rectify wrongs wrought against people by local bullies and clumsily armored, horse-mounted landowners.

Arms, armor, precious stones, skins, furs, fabrics, spices, incense, hand-looms, and other hand-tools were the principal goods traded in Roland's time. Gold, silver, and pewter served as money. Trading was accomplished on foot, on the backs of animals, or on river-borne small craft. The land of the overlord was the principal wealth.

Squads of armed horsemen could protect caravans of goods-carrying horses, camels, and elephants along with human bearers. These caravans could transport the initially culture-evolved riches of the Orient westward to the ever more westwardly advancing frontiers of humanity, where the newly powerful cultures could acquire the historically recognized appurtenances of Oriental courts of power.

A new kind of wealth-making occurs historically with the invention and development of stoutly and heavily keeled, ribbed, and planked, high-seas-keeping, deep-bellied, and, in much later times, cannon-armed sailing ships.

These great ships were built in vertical shorings. Their keels were laid upon heavy wooden cross-ties and blocked against premature sliding. These cross-tie "ways" led down very gently sloped banks into the harbor's deep waters. When the ships' hull was completed and watertight, the cross-tie ways were greased and the blockings mauled out from under the ship. Gravity slid the ship swiftly seaward, maintaining its vertical balance long enough to plunge it deck-side-skyward into the water.

After launching, the ship was floated progressively into a succession of wooden crane-equipped outfitting docks— the interior decks and bulkheading dock; the chain-plating dock; the mast-stepping dock; the rigging and sail-bending-on dock; the winch-, capstan-, and armaments-installing dock. Finally she sailed away to various lands where superior masts, fabrics, ropes, etc., progressively replaced the original make-do equipment. (World's best masts from the Pacific Coast of British Columbia; best rope-making fiber from Manila, in the Philippine Islands; best cotton fiber for the sails from Egypt; best teakwood for the decks from Thailand.) It took complete circumnavigation to incorporate the "best in the world" of everything to produce a "gallant" ship—one capable of around-the-world sailing.

It is probable that the first moving-line shipyard in history was established on the Chao Phraya River in Bangkok. However, the earliest now known militarily secure shipyard is to be found on the Greek island of Milos. It is in a miniature rock-walled fjord, well hidden from enemies by a deep-channeled, curved entrance. On the many rock platforms lining the fjord's walls, many shipbuilding artifacts were found. The Milos shipbuilding fjord was so well hidden that the Germans used it for their Aegean Sea submarine hideaway during World War II. (The *Venus de Milo,* now in the Louvre in Paris, came from Milos.)

History's next great moving-line shipyard is as yet to be found in Venice. So strategically important was the Venice shipyard that it was initially seized by Napoleon early in his campaigning.

Centuries later this progressively moved-forward-and-added-to shipbuilding pattern as yet clearly evidenced in the Venice shipyard became the prototype for all of mass-production industry's "moving lines."

The ship was, of course, a tool, but not a craft tool produced by one man. It was an industrial tool, mass-producible and operable only by large numbers of highly skilled craftsmen, metalworkers, woodworkers, sailcloth-makers, rope-makers, iron chain- and anchor-makers, seasoned sailors, and the coordinated muscle of "all

hands." The merchant ship was a wind-energized industry, a tool that could sail around the world and carry
cargoes worth many fortunes to lands not containing the
materials brought by the ships, which when integrated
with the home-port-occurring materials produced real
wealth of increased life-support for more and more people.

The building, rigging, and arming of such vessels and
the production of the materials with which to build them,
as well as the production of the food and other necessities
to feed and clothe all those engaged in the shipbuilding,
required an effectively powerful military authority able to
command the full-time commitment of the work and skills
of the large numbers of humans involved. It also called for
the amassing of large sums of negotiable wealth. Prefera-
bly the negotiable wealth was in the form of trade-imple-
menting precious metals and jewels, commercially accept-
able around the world.

For ages earlier the negotiable wealth had been the
efficiently demonstrable products of labor and its produce,
the grains and the livestock. Of the latter, the protein-
amassed cattle constituted the most concentrated possible
yet maneuverable realization of actual life-support wealth.
Cattle were put up as collateral for the banker's loan of
gold, silver, and copper coinage. When the voyage was
successfully completed, the merchant-ship venturers repaid the banker and paid the banker his "interest" in the
form of calves that had been interimly produced by the
collateraled cattle. This was called "payment in kind"—
kind being the kinder or "children" of the cattle. When
bankers eliminated live cattle as collateral and dealt only
in gold or silver, there were no gold coins being bred by
gold coins as calves had been by cows, so interest was
taken out of the capital gold by diminishing the equity of
the borrower when he repaid his debt. The banker's inter-

est was cut out of—that is, deducted from—the depositor's original "cap"-ital (head of cattle) stake.

As I made clear in *Operating Manual for Spaceship Earth,* * when the farmer or cattlemen producers of "real wealth" of one hundred forward days of life support each for one hundred people—i.e., one thousand man-days of life-support—deposited their monetary specie equivalent in the bank and the banker loaned it out at 10 percent, it meant that the banker stole one hundred man-days of life support from the farmer depositor instead of providing the farmers the bank-advertised "safekeeping." The banker could hide this situation by price increase in the profits the banks made by using the depositor's real wealth units. But the depositor's dollar could buy him ever less real life-support units.

The safe return of the merchant venturer's ships was so unpredictable as to constitute a capital investment of high risk but also of very high potential gain—most significantly a risk whose rewarding payoff might take several "crop" seasons to realize. The voyage might take several or even many years to complete. These risks in turn could be lessened by insurance.

As a consequence of all the foregoing, a half-millennium after Roland a new and overwhelmingly greater form of invisible seagoing and land-strutting giants appeared on planet Earth. This was a legally contrived, abstract giant —"legal" because the physically uncontradictable "top-sword" king *decreed* it was legal. Having the most favored privileges accorded real humans, the giant, abstract, corporate "man" is inventively created in 1390 in England. (The corporate "human" may have been invented in ancient Babylon to cover the potentates' voyaging ventures,

---

*Published by E. P. Dutton, 1969.

but we have as yet no written record of such.) "His" abstract name is the "Merchant Venturers Society." This composite man was formed by the king of England with a small group of his very powerful friends, who lorded over their king-deeded vastlands.

By royal prerogative, the venture-financing riskers could not be held liable for any losses of the venture. With limited liability, individuals might sue the company but not the human individuals who underwrote the venture. If the enterprise failed and went bankrupt, its shareholders lost their ventured stake but were not to be held responsible in any way for its debts. The creditors of the company were the losers, and not the shareholders. Bankruptcy could reflect no credit stigma upon the companies' shareholders. The shareholders were held absolutely blameless for any misfortunes of their ships' crew or for damage caused by collision of their ship with another ship. If the ship and its cargo were lost, the shareholders lost their original shares, but no more. As long as the ship operated successfully, the shareholders shared its trading profits.

Whether the ship was lost or not, the banker who loaned the gold for the merchant ship's trading held the life-support-producing lands and their cattle as collateral. Since many voyages ended in disaster, the banker occupied a long-time, steadily profitable position in the overall merchant venturing—and as yet does.

Naturally, the shareholder's limited-liability advantage, granted by sovereign decree, encouraged a swift expansion of such enterprises.

In 1522 Magellan's ship demonstrated that the world is not a laterally extended plane off the edge of which a ship might plunge, nor an ocean extended laterally to infinity from which there was no return. Magellan's ship's circumvoyaging proved that the Earth is a sphere—a closed system with enormous trade-monopolizing potentials. Laws of the land could not be enforced on the sea. The seagoers were outlaws—privateers or pirateers. The most powerful outlaws became the sovereigns of the ocean sea.

In 1580, Queen Elizabeth was the largest shareholder in Sir Francis Drake's merchant ship *The Golden Hind.* Naturally, the queen granted Drake's venture "legal" freedom from liability. After paying Elizabeth her conspicuously major share, Drake and his other shareholders each realized almost 5,000 percent profit on their risked capital.

Enthusiastic over her *Golden Hind* venture, in 1600 Queen Elizabeth chartered the limited-liability East India Company. This time the shareholders acquired shares in a fleet of ships, docks, and warehouses in both England and India—not shares in just one ship, as in the earlier "venturing."

Employing her sovereign power, Elizabeth limited the losses of its chartered riskers to their initial monetary or equivalent capital stakes, while continuing their right to receive their proportional profit dividends for as long as the venturing company might exist—in perpetuity.

Known later in England as "Ltd." (for "limited liability"), in France as "Société en Commandite," in Germany as "Kommanditgesellschaft," and as "Corporation" under the U.S.A.'s "Inc." (for incorporated) status, this newborn abstract legal giant was to be treated as a human personality, empowered to do anything humans can do but also accredited to operate as an abstract, legal entity able to enter or leave any nation without a passport. As such it was able to employ millions of people and any amount of money, tools, buildings, and equipment, and to perform its giant acts anywhere about the oceanic world exclusively for the profit in perpetuity only of its shareholders.

When the Fourteenth Amendment to the U.S.A. Constitution was passed in the post–Civil War railroad-expansion days, the U.S. Supreme Court required that the individual states grant the corporation all the privileges and protection granted to human citizens. A hundred years later, in 1980, the U.S. Supreme Court ruled that a corporation had the same rights of free speech as all U.S. citizens.

To allow its corporate bodies to make a colossal new grab, Grunch has ordered its pet puppets to take over the world ocean-bottom resources. As of February 1982, the United States, Britain, France, and West Germany have reached preliminary agreement to bypass the stalled Law of the Sea Conference and proceed with development of seabed mineral resources, the Japanese foreign ministry said. Japan expressed opposition to the agreement—unconfirmed by the four other countries—and said such a program should operate under U.N. auspices. The United States and other developed countries have refused to agree to developing nations' demands that seabed development be overseen by a U.N. agency dominated by the poorer countries.

The fourteenth-, fifteenth-, and sixteenth-century rulers who instituted and empowered those abstract corporate giants were able to popularize their acts by celebrating the visual wealth of goods it brought to their country and to the political satisfaction of their many citizens. The profit to society was *visibly* distributed as the goods, services, museums, and public-place rarities the enterprising produced. The shareholders' dividend checks were invisibly distributed.

With the battle of Trafalgar in 1805, the risk-capital powers backing the "British Empire" became the "Sovereigns of the Seas." Until that time the high-sea venturers had carried gold and silver as their trading medium. This induced world-around high-seas piracy. The behind-the-scenes masters of the British Empire then invented the "annual balance of trade" as a world-around bookkeeping system which kept its gold off the seas and instead, after the year-end tallying of the trade interactions, transferred the gold from one country's London vault to another country's London vault. This withdrew the gold from the seagoing pirate's reach. However, it brought many of the pirates into the financial districts of great cities.

Naturally, shareholding in Ltd. enterprise became increasingly attractive as an investment risk, but soon the monetary size of investment required for share participation grew beyond the acquisition means of all but the wealthy. Stock-exchange brokers, for their own convenience, imposed trading only in hundred-share "lots" or "blocks," which quickly raised the equity-purchasing increments to so great a price that only the very wealthy could any longer participate in such venture-sharing. The capital games' playing-rules "kept the pikers out," the original pikers being the on-foot, pike-bearing castle guards.

In the nineteenth century the limited-liability corporate venturing began not only putting its shipyard donkey engines' steam engines in ships, but also mounting them on steel wheels on rails and powering them out of the shipyards. Thus they began railroading heavy loads inland. This initiated new mass-production industry centers at inland water-power sites. For instance, industrial venturing underwrote water-wheel-driven mass-yardage cotton-mill fabric production, perferably in such low-wage-paying countries as India. The annual balance-of-trade accounting brought about many obviously inequitable economic conditions, such as, for instance, India's burlap-bag-makers working for a penny a day. It was the vast profits made on burlap bags so produced which financed the early-twentieth-century expansion of the Massachussetts Institute of Technology in the U.S.A.

Such cotton and woollen fabric production-venturing was logically followed by thread and needles, pins, buttons, and small hardware mass-production moving-line ventures. With the introduction of electricity and the electricity-driven motor, industry began moving-line mass-production of dollar watches, tin cans, safety razor blades, big-city clothing-production sweatshops, then bicycles, then motor cars. In World War I, it introduced steel steamship mass-production; in World War II, transoceanic aircraft mass-production; and, in the "cold," puppet-nation-waged war (World War III), extraterrestrial travel and transport, and mass-production of invisible mass-killingry potential.

•   •   •

There is a fundamental evolutionary patterning in which, with each new era and phase of technology and social-economic venturing, both the tools and their products get bigger and bigger, and the numbers of humans involved multiplies. A period of doing more with more until a mammoth peak magnitude is attained which is followed by evolutionary production of ever more effective results with ever less pounds of material, ergs of energy, and seconds of time, all of which integrating synergy produces ever more comprehensively effective tools with ever smaller technological artifacts produced by ever fewer unskilled human workers—the 1895 to 1929 model "T" waxing to the 1960s Cadillac limo, then waning to the 1980 Japanese Honda.

For example, trans-ocean traffic brought into use ever more gargantuan ocean liners leading eventually to the five-day-Atlantic-crossing leviathans, such as the 81,000-ton *Queen Mary* and her sister ship the *Queen Elizabeth.* Using the World War II technology's new, lightweight, high-strength, saltwater-impervious aluminum alloys in her superstructure, the *S.S. United States* was built to carry the same number of passengers and the same amount of cargo, and to cross the Atlantic at the same speed as the *Queen Mary,* though weighing only forty-five thousand tons, that is, 55 percent of the weight of the Queens.

These five-day-Atlantic-crossing passenger carriers are now obsolete. In 1961, three jet airplanes outperformed the *S.S. United States* in carrying capacity, in hours instead of days and at less expense.

In 1980, ever lighter, swifter "liner"-type steamships are being built, but only for luxury cruise ships. For twenty years, these obsolete ocean liners have been progressively replaced by ten-to-thirty ton, one-third-of-a-day-transatlantic-crossing jet aircraft.

Another example of the little-to-big-to-little evolution is manifest in the world of mathematical computing. In developing trigonometry and its solution by logarithms, thousand of monks worked for hundreds of years to produce the one-degree tables of sines, cosines, tangents, and cotangents. During the Great Depression years of 1930 to 1936 the British and German mathematicians were hired by their governments in a joint project to calculate the table of functions to a one-minute of arc exactitude. Then came the big post–World War II calculating machines, Univac et al., filling whole university buildings with thousands of thermionic tubes. Then came the tubeless transistor and computers weighing and bulking far less, until we came to printed circuits and "chips" and table-top equipment doing better work than the whole-building-filling equipment. Before all this, I myself spent two pre-calculator or -computer years carrying out the trigonometric calculations for geodesic domes. I had to do so "longhand." Then appeared seventy-five-pound electric calculating machines, followed by the pocket-size computers with which the trigonometric problems that took me two years of work became solvable in one day by one person.

This process promises within a few years to become so miniaturized and so comprehensively capable as to be the size of a hearing aid, though able to interact with all the world-around computers and able to discern how best to operate our planet, making obsolete the opinions of corporate or government executives.

•   •   •

As mass-capital-venturing flourished after World War I, General Foods Company absorbed many pre–World War I individually owned, independent mass-producers of canned and packaged food. General Electric acquired other successful electrical goods manufacturers. The growth of corporate venture activity was, however, at that time yet identified by unique product categories.

After World War II, "mergers and acquisitions" and outright "takeovers" agglomerated almost all successful industrial capital ventures, regardless of their class of produced goods and services. The great conglomerates found it more profitable, safer, and more credit-powerful to diversify their risking. The successful "biggies" became ever more gargantuan—for example, the Dupont chemical company's 1981 acquisition of Conoco, America's ninth-largest oil company, for $7.57 billion, to form the seventh-largest industrial corporation in the U.S. Because many of these conglomerations embraced all the national defense weaponry production, they "legally" qualified for guaranteed government "bailout" should their operation become financially "embarrassed" or debts unmeetable. The U.S. government's decade-ago bailout of Lockheed Aircraft or its multibillion-dollar guaranteed loan to private Chrysler Corporation (the government's military-tank producer) are the current outstanding examples.

Chapter 4

Invisible
Know-How, Inc.

As mentioned earlier, limited-liability, abstract corporate "beings" needed no passports to travel altogether invisibly across national borders. Soon after World War II, America's five hundred largest corporations became supranational, taking with them (out of the United States) the invisible legal controls over what had been born as American industry with all its "know-how." The know-how had been paid for initially by the U.S. people through their government's wartime (or "on the brink of wartimes") underwriting of the prime technologies as initially developed only for the U.S. Department of Defense or the Manhattan Project or the space program, developed in wartime at government ("we the people's") expense and turned over gratis for "operational efficiency" in "peacetime" to privately owned corporations.

World War I brought vast munitions-buying on credit by the U.S. government, and the figures ran into multi-millions of dollars as private U.S. industrial corporations acquired postwar operational rights to all the wartime government-financed new-era technology production machinery. Stockholders prospered. World War II saw the same U.S. government credit employed to produce "multi-vaster" new-technology munitions, with the dollar figures running this time into the multi-*billions* of dollars. World War III's third-of-a-century of "cold-warring" between the U.S.A. and U.S.S.R., waged vicariously through many hot-war puppeted nations, has seen the annual munitions figures running into the multi-*trillions* of dollars. The U.S.A. 1981 "national" debt is over a trillion dollars, and the U.S. cannot pay even the interest on that debt. We can very properly call World War I the million-dollar war and World War II the billion-dollar war and World War III the trillion-dollar war.

In the meantime, all the industrial research and development as well as its products have become involved with the invisible technologies of atomics, electronics, chemistry, molecular alloying, and information processing. All the research and development of all the products and services that are going to affect all of our forward days are now being conducted in the realm of the electromagnetic spectrum "reality" not directly apprehendible by any of the human senses.

While the North American–situated factories and spectacular city buildings seem to be and are thought of by humans as being American property because they are located on American land, most are no longer U.S.A.-people-owned. For instance, though thought of as "American," a majority of the skyscrapers of Honolulu belong to Japanese bankers. Arabian oil billionaires own many U.S. city skyscrapers. Kuwait owns the large South Carolina coastal island of Kiawah.

What was once world-around high credit for American

ingenuity and friendliness is no longer existent. On February 1, 1982, the United States ambassador to the United Nations stated to the media that all the "United Nations now hate the U.S.A." What they hate is Grunch, but Grunch is able to deceive the world into blaming the very innocent people of the United States.

All the continental U.S.A.'s industrial factories and grounds and 90 percent of all that can and does produce physical wealth has already become or is about to become the humanly invisible property of inhumanly operative supranational corporations controlled by the invisible human owners of invisible Swiss bank account code numbers.

A vast new giant of approximately no-risk capitalism is now astride the world. "Earning" over a trillion dollars a year, this supranational giant's monopoly over know-how, wealth, research and development, and production and distribution facilities is worth at least $20 trillion (U.S.A. dollars, September 1981). While the giant now owns and controls four-fifths of the planet Earth's open-market bankable assets, $1 trillion of those giant's assets are in monetary gold bullion. Astride spherical Spaceship Earth, the supranational corporate Grunch of Giants faces a political giant of noncapitalistic forces controlling the lives of two-thirds of humanity.

In making these observations in regard to inanimate corporations we do not infer antisocial attitudes on the part of the corporate officers. A corporation's executives are elected by its board of directors. The directors are elected by the number of shares of stock as voted directly by their holders or as voted by the holders of their share's proxies. This voting is not on a democratic one stockholder/one vote basis but on an as-many-votes-as-shares-owned basis. This being so, the corporations' lawyers have no alternative to reminding any altruistic, socially concerned executives that the corporation is committed by law only to making money for its shareholders, and therefore that any socially concerned, altruistic proclivities of any corporate executive must be realized outside the corporation and at the executive's own expense.*

In the August 3, 1981, issue of *Time* appeared the following article:

President [Reagan] appointed William Baxter, a Stanford law professor who firmly believes in the virtues of large-scale enterprises unfettered by excessive Government regulation, to be his antitrust chief in the Justice Department. Baxter's boss, Attorney General William French Smith, succinctly stated the new Administration's philosophy in an oft-quoted speech before the District of Columbia Bar. Said Smith: "Bigness in business is not necessarily badness. Efficient firms should not be hobbled under the guise of antitrust enforcement."

Baxter openly accepts some responsibility for the merger phenomenon. Said he last week: "The statements we've made at the Justice Department have allowed people to think about mergers that they really wouldn't have thought about in past Administrations." Mobil's bid for Conoco is

---

*For all the same basic reasons, the inanimate, literally soulless and heartless corporations cannot feel and express human sensitivities and thoughts, such as I find printed on desk-top cards in my hotel rooms around the world. I find it specious for a hotel chain to assume the role of moral arbiter by, for instance, printing cards displayed in their hotel rooms which define "love" as being the act of forbearance from stealing the hotel's towels or by exploiting the public concern over energy problems by asserting on their room cards that "love is saving the electric current costs." Since the corporation is only a legal device, the only possible reason for paying those "love" cards' printing costs is to reduce the hotels' operating costs, thus hopefully to increase the corporate dividends. Such operational tricks may well bring about promotion for the "ingenious" executives who conceive them.

a case in point. Such a merger between two of the top ten
petroleum companies would never have been seriously con-
sidered during Jimmy Carter's term. Baxter insists that his
trustbusters will not allow any acquisition that significantly
reduces competition within the oil industry or any other.
He also maintains that a Mobil-Conoco combination would
be subjected to tough scrutiny in Washington. [That is one
reason why the subsequent alternative deal which united
non-oil Dupont and oil Conoco was countenanced—
R.B.F.]

Baxter should be wary if only because the American
public has long been apprehensive about excessive corporate power. [Attorney General Smith] admits, "The strains
of populist hostility toward large companies are deeply
ingrained in the U.S.A. Government trustbusters have enjoyed broad public support as they attacked both concen-
tration within an industry and combinations between corporate giants in unrelated business." Yet the burgeoning
growth of corporate America has outpaced all the antitrust
efforts. Since World War II, the portion of U.S. industry
controlled by the 200 largest manufacturing firms has risen
from 45% to 60%. [Socioeconomically, that is from majority to minority control—R.B.F.]

The attorney general chooses his words carefully. What
he speaks of as U.S. industry is not the ownership of the
corporations conducting the industrial activity; he speaks
exclusively of the physical production activity itself taking
place under the roofs of factory buildings situated within
the geographical borders of the U.S. of North America.
The capital title to and productive earnings of these are
60-percent owned and controlled by the entirely unknown
majority owners of the escaped-from-America, supranational corporations—the Grunch.

One-third of humanity lives outside the lands controlled
by socialism. All unbeknownst to and undetected by the
one-fifth of the one-third of humanity residenced within

the U.S.A., gradually cross-breeding "worldians," their one-third-of-a-century-ago kudos for realistically articulated generosity to and concern for others, as well as the U.S. peoples' legal ownership and control of their economic assets, have been altogether exploited, usurped, or stolen from them by the invisibly integrated supranational corporate giants. The Grunch has conducted its ruthlessly selfish activity always in the name of the U.S.A. people.

• • •

The now majorly literate crossbreeding world humans are now looking askance at both the socialist and capitalist giants as these politically opposed powers multiply their to-anywhere-deliverable, humanity-annihilating bombs.

"Modern weapons are growing so sophisticated and so small [*ephemeralization*—R.B.F.] that any future arms-control agreement would be impossible to monitor and enforce," according to a Knight-Ridder News Service dispatch. "What we are going to *see* ["experience"—R.B.F.] in the next generation of weapons is *invisibility,* which translates into insecurity," according to William Kincade, a former naval intelligence officer.

Even such an informed source as Admiral Stansfield Turner, former head of the grand-naval-strategy-formulating U.S. Naval War College in Newport, Rhode Island, then commander-in-chief of the U.S. Navy's Mediterranean fleet, and then Director of the Central Intelligence Agency, says: "Any hope of limiting total destructiveness is slipping past us."

In the affairs of the supranational corporate giants, real quality of product, consciously sustained, has given way to packaging-allure and advertising-proclaimed "quality" as commensurate only with the best interests of corporate moneymaking. As already mentioned, heads of great corporations are elected by the stockholders' directors, who in turn are chosen by those controlling the majority of voting shares, who make their choices only on the basis of greatest earnings performance. Operating only as abstract, global-magnitude legal entities, all the unknown-owned-and-controlled supranational corporations have no human-community consideration other than as potential customers, consumers, or fighting-force conscriptees.

At the termination of his presidency, Eisenhower expressed his shocked dismay over the exclusively self-concerned military-industrial complex that he had found to be growing inexorably as a malignant economic organism. There is no question of Eisenhower's innocence of such a phenomena as he assumed his great responsibility. In the same way I am confident that Reagan is utterly unaware of the existence, magnitude, and nature of the supranational colossus. He knows he is dealing with rich and business-wise-proven individuals whose organizational management effectiveness is of a high order. Because the colossus is operating an invisible technology, and society is so specialized that each individual is acquainted with only a few of the billions of other specialized invisibilities, and because of the invisibility of who the supranational shareholders may really be and where they are, I am confident that Reagan truly thinks that he is operating strictly within the historic limits of a U.S.A. national government and not as a stooge of an invisible Grunch of literally soulless supranational giants. I don't think David Rockefeller or any of the justices of the U.S. Supreme Court, or Volcker, head of the Federal Reserve Bank, or Margaret Thatcher, or the heads of any of the world's governments think of their problems in the realistic terms of their being governed entirely by the inanimate, socially unconcerned, supranational colossi, as the possibly lethally nonhuman growth could prove to be. However,

Grunch could also prove to be an army of benign giants, because it will depend more and more on its complex, world-around computerization integration, and the data entered into that integrated network will continually evidence that the present technology could make the world work for everyone, and at a much more profitable level than realized from weapons production. Nor do I think these present power structure spokespeople see the supranational corporations as the unwittingly benevolent agent of evolution about to close the historical era of separate "nation-states" and to institute in its stead the era of omni-economically successful, omni-integrated planetary society.

For the past thirty years the U.S. government's grand defense strategy has been a puppet of the supranational corporate giants with the strings invisibly manipulated. This policy has concentrated on accumulating greater numbers of atomic bombs than those of the U.S.S.R., while all the while the U.S.S.R. was (only) ostensibly endeavoring to keep pace with the U.S.A. bomb production. In reality the U.S.S.R. was dominantly preoccupied with building an all-oceans, primarily underwater navy from scratch and expanding numerically its conventional-weapons army divisions.

At the time Eisenhower became president of the United States, the military experts of both the U.S.A. and U.S.S.R. had independently concluded that a missile-delivered atomic war would be the first war in history in which both sides would be utterly devastated. In gun-munitioned warfare, whoever shot first and accurately won. The other man's shot never got away. In 14,000-miles-per-hour delivery rocketry warfare, the 670,000,000-miles-per-hour operating radar vision of both sides gives each side enough advance notice after its respective enemy has fired to let loose all of its arsenal before the enemy's missiles arrive. For the first time in warfare history, both sides utterly lose. For those hotheads in the capitalist world who as yet contemplate pre-emptive firing of the U.S.A. arsenal of atomic warheads, it is importantly relevant that the Russians have accurate, geographically triangulated positioning of their U.S.A. targets, while the U.S.A. does not have accurate geodetic triangulation of the location of most of the U.S.S.R. targets.*

• • •

To best understand the present (November 1981) world crisis, it is necessary to turn history back for almost a century, back to when Edison invented the electric lamp and the direct current generator. J. P. Morgan, Sr., the economic power structure giant, was the first to act upon the realization that: whoever developed, manufactured, installed, and controlled the physical-energy generators and the metered-energy distribution and cut-off system could and would control the national economies into which they were physically introduced. The air we breath was everywhere so plentiful that its availability could not readily be monopolized. There were too many ponds, lakes, rivers, brooks, and wells to make the metered water-supply systems a generally monopolizable business.

When Alexander Graham Bell invented the telephone, it had to compete with the post-office conducted mail and required far greater numbers of employees. Morgan saw that the copper mines and the electric equipment manufactured from copper as well as all the power-generating companies involved the least labor participation and the then maximally profitable business.

All of the foregoing required the availability and con-

---

*See *Critical Path,* "Triangulation Mapping," pp. 184–188.

trollability of an utterly unprecedented magnitude of physical apparatus and installation of otherwise unemployed monetary wealth. The patents of Edison's inventions and an army of astute lawyers and brokerage houses became the pivotal legal-precedent-accepted economic properties and work force in amassing the initial procurement capital of Morgan's power monopoly.

This initial capital-amassing was greatly augmented by selling interest-bearing bonds to widows and trust funds in general, seemingly safely secured by the vast lands given as a "grateful" U.S.A. people's government as a subsidy to the pioneer railway-building and -operating companies. The railroad company bonds were secured by the seemingly highly valuable real estate adjacent to all the railroad, cross-country rights-of-way, their way-station town properties, railway stations, trackage, etc. which railroad company bonds were purchased for widows and trust funds in general by their trustees. This capital amassing initially financed the electric power companies. As we have noted elsewhere, these railroad bonds became worthless in the 1929 economic crash. Nobody wished to buy the old depot buildings, etc.

When the automobiles and the auto-trucks took so much business from the railways as to render the railroad passenger systems profitless, and cross-country, pipeable petroleum replaced coal as a prime fuel, and giant trans-oceanic tankers were developed, and the Middle East oil lands were explored and developed, the petroleum business rose swiftly to become the maximum economic power giant of the twentieth century, outpowering the Morgan utilities- and banks-based system.

The number of kilowatts of electric energy being generated from each BTU (British Thermal Unit) of fossil fuels burned or foot-pounds per second of water-power-derived turbine-functioning has continually increased since the very beginnings of electric-power generation and distribution. Concurrently, the weight of the production and distribution equipment to produce that power rapidly decreased per each kilowatt or horsepower of energy produced and delivered. As a consequence of this never-ceasing technological increase in overall efficiency, the actual overall cosmically predicated costs of energy generation have always and only decreased, and cost increases have been the consequence of those in top power-positions contriving through pricing to be able to pay ever greater dividends to shareholders and thus to increase the stock-market value of their own shares, thus in turn to increase their power to control the amassed money of others as capital. Such capital power manipulation is intoxicating and seemingly unchallengeable.

However, as with all socioeconomic-political power evolution, the politically appointed "public service" commissions in all the states have consistently granted even higher kilowatt-hour price rates to the privately owned, deceptively named "public service companies" producing the electric power. So unchallengeably powerful are the "public service" commissions that in 1981 and 1982 they have been able to allow great utility companies to abandon some nine hundred million dollars wasted by utility companies as they abandon their partially finished atomic-energy plants on the U.S. West Coast—charging the loss to the consumers by increasing their rates. This results in private enterprise making a $900 million bad gamble and having the capability of passing on their loss to the public.

The constant fundamental operating-cost reductions, combined with constant price increases, have produced so much money that the power-generating businesses are amongst the wealthiest and most invisible of the politically manipulative organizations. After World War II, the electric power industry's three-quarters-of-a-century-accumulated wealth successfully combined its political power with that of the oil giants to "take over for nothing" the total atomic-energy program assets. This included all of the know-how and production apparatus of the U.S. government's military atomic-energy program, for which development the U.S. citizens had paid $150 billion.

This amassing of political power coincided with a generally dawning awareness of U.S. youth in general and an ever-increasing percentage of the mature U.S. electorate regarding the corruption of the political representatives of their theretofore-trusted democratic government. This corruptibility is inherent in the fact that the TV electronic campaign costs of U.S.A. elections now amount to $50 million for the presidency, $10 for a senatorship, and $5 million for a congressional seat. This corruptability is enhanced by the U.S. Supreme Court's hang-fire no-ruling of 1981 which will allow unlimited money to be spent in the next election years.

The now-gone supranational corporate giants have always known that the fossil fuels can and will become exhausted. To meet this contingency, their post–World War II last-third-of-a-century grand strategy has been to force the U.S. government to develop superior atomic-war capability, knowing full well that atomic warfare will terminate human occupancy of planet Earth, which fact would eventually force the government to abandon its war use. First and foremost, however, the power monopoly would have to have accomplished their "public service" atomic energy objective; first, of becoming contractors to operate government atomic facilities; second, of siphoning off from the U.S. government all the latter's atomic scientist personnel and all the invisible know-how to develop world-around atomic-energy plants to feed into their wired and metered energy-monopoly system as the petroleum source diminished and approached depletion. In the meantime, while maintaining their power over the U.S.A. and other political systems, their grand strategy found it

necessary to have the U.S.A. and Western World population satisfied that the U.S.A. was successfully maintaining
its fighting superiority over the U.S.S.R. by producing
more atomic bombs than the Russians.

•   •   •

To initiate his wired and metered electric-energy-power
monopoly in the "gay nineties"—1890—which three-
fourths of a century later became an overwhelming socioeconomic power, the elder J. P. Morgan used the earlier
formula of issuing bonds and preferred stocks on each of
his enterprises as soon as they were paying dividends. He
was thus provided with additional free capital to initiate
other branches of the power-structure system: for example, in copper mining (for use in the generation and con-
duction of electric power), steel manufacture (for the high-
line masts and structural housings of the electric
equipment), etc.

He used his engineering firm of Stone-Webster to design
and build his foreign-country power systems operated by
Electric Bond and Share Company—EBASCO. His price-
increasing by the power companies was automatically
matched by increase in the stock-market sale of his companies' shares. These share values increased with his own
equities' advance. Using these equities as capital, he
opened his own banks.

Because his enterprising monopolies earned good dividends, shares in his companies became increasingly popular. His own bank and the banks he controlled opened
brokerage departments. He backed the opening of many
stock-exchange-seat-owning   individuals'   brokerage
houses to cope with the increasing complexity of open-
market selling and buying of his companies' shares or
bonds. By the time of the 1929 Crash, Morgan was controlling the boards of directors of General Electric, General Motors, U.S. Steel, the big three copper companies,

the telephone and telegraph companies, all the "Edison Electric" public utilities, etc.; and many of the U.S. banks.

At the outset, Morgan's partners gave Harvard University its law and business schools, from whose highly educated, specialized graduates they recruited the army of lawyers and financial experts to service their Wall Street offices. This legal army handled the behind-the-scenes complex contractings and financial paperwork implementing Morgan's and his associates' enterprises. There being no laws against so doing prior to 1929, he used general bank deposits to underwrite his enterprises.

In the early 1920s, the Morgan-dominated banking system pushed farm machinery sales to farmers on time-payment plans secured to the banks by first mortgages on the farm properties as well as on the machinery. As I explained in *Critical Path,* the bad hog market of 1926 hit farmers financially, causing many to be unable to make their monthly payments on their time-purchased farm machinery. The country banks not only replevined the machinery but foreclosed on the farms, which were mortgaged to guarantee the time payments—the country banks found the farms unsalable, as there were no other U.S.A. individuals eager to go into farming. ("How You Gonna Keep 'Em Down on the Farm After They've Seen Paree?" —World War I song.) Then the bigger city banks, which had loaned the small banks money based on the "soundness of physical land and machine collateral," foreclosed on the small country banks. The larger city banks also found their foreclosure properties unsalable. No cash funds were available to accommodate their depositors' withdrawals. "Runs" on banks multiplied. There came a crisis moment when over five thousand banks closed in one day. Finally the big Chicago banks closed and only the big New York banks remained open. Then it was discovered that they, too, having loaned their deposits for industrial ventures, now lacked cash monies with which to refund their depositers and the New Deal and FDR declared the "Bank Moratorium," thereby avoiding admitting the bankruptcy of the U.S. banking system and with it the end of U.S.A. capitalism. The U.S. Congress, inquiring exhaustively into the matter, found that those New York banks' brokerage departments had been using deposits for underwriting venture industries. Because this was at the heart of the failure, the Glass-Steagel Banking Act of 1933 was enacted "permanently"—it was hoped—separating venture brokerage-underwriting from the New Deal government's guaranteed bank deposits of the people.

J. P. Morgan, Sr.'s strategy of organization of the financing and control of U.S.A. industrial development was evolved from his close cooperation with the Bank of England's centuries-old, behind-the-scenes laws of accepted precedent of physical property rights and their convertability into "paper securities" as marketable shares in enterprise. The Bank of England and Morgan nurtured the young Grunch of eighteenth-century giants from their youth into lusty nineteenth-century colonial maturity. J. P. Morgan became the official fiscal agent of the British Empire in World War I. As the "Allies' " purchasing agent from 1914 to 1918, Morgan's amassing in the U.S.A. of the profits of World War I shifted the world capital of the grunch of corporate giants from Europe to America, and until the 1929 crash assured Morgan's dominance of the socioeconomic evolutionary balance of power over human affairs on planet Earth.

•   •   •

The Russian Revolution brought about the 1917 inception of the U.S.S.R. and its organization of world communism as an evolutionary challenge to capitalism's power.

In *Critical Path* I have traced these evolutionary events to 1981. In 1981, the supranational invisible moneymaking colossus finds itself faced with the U.S.S.R.'s 1981 superiority in number of conventionally armed divisions and greater all-oceans naval power than that possessed by the U.S.A. and its NATO allies.

U.S. Secretary of Defense Caspar Weinberger admitted in April 1981 the extraordinary Soviet military buildup, which has left in its wake the U.S. loss of strategic defense advantage which it had maintained in the fifties and sixties. He cites Defense Department data showing the Soviets' four-to-one advantage in tanks, their two-to-one tactical advantage in atomic-powered, atomic-missile-launching submarines, and a four-to-one advantage in submarines, and in addition the sizable increase in the weight and accuracy of their nuclear intercontinental strategic missiles. He describes their naval buildup as the fastest in naval history. Weinberger states that, at present, the U.S.S.R. continues to outbuild the U.S. two-to-one in surface vessels and five-to-one in submarines, according to his latest figures. (*Jane's* annual publications of international armaments have been providing these figures quite accurately over the years with no mention of them by the petro-atomic-power-structure-puppeted U.S. politicians and even less mentioned by the supranational-giants-controlled U.S. press.)

To regain its military edge (detente) over the Soviet Union, if it can be done at all, and assuming that the U.S.S.R. does nothing to offset the attempt, will take ten years and, according to Defense Secretary Weinberger, $1.6 trillion or approximately $1 billion a day ($365 billion a year) over the next five years, just for starters. The supranational-money colossus needs that time to build its CIA-organized and U.S. people's unknowingly financed fighting power, officered by mercenaries* and soldiered,

---

*For example, the aborted Seychelles "invasion" of November 27, 1981, involving mercenaries based in South Africa.

sailored, and piloted by puppeted countries to attain a superior posture over the Soviet Union or any other power —including even a possibly-to-be-catalyzed, supranational, individually thinking and acting, spontaneously cooperating amalgam of now majorly literate, apolitical, world-around citizenry.

Since it takes millions of dollars to win U.S. elections, the vast majority of America's crossbreeding youth and an ever-increasing number of its adults concede politics to be so inherently corrupt as to cause increasing numbers of qualified voters to withhold from voting, lest in doing so their action be misconstrued as constituting their approval or acceptance of the present-day corruptability of politics and its consequent inability to articulate the will of democracy.

In 1980, of the 227 million persons in the United States (159 million of whom were eligible to vote), only 78 million, or only one-half, of the eligible persons voted in the most negatively momentous presidential election in nearly a third of a century. Of that number who voted, only 40 million voted for the winning candidate, Ronald Reagan. The "overwhelming majority" that President Reagan repeatedly claims legitimizes his "mandate" for sweeping executive and legislative change consists, in fact, of only 14 percent of the people of the United States.

Many youth and many oldsters inspired only by a concern for all Earthians and convinced that their voting cannot stem big money's "will" did not vote. On the other hand, the economically rich, seeking to secure their economic advantage, for example rich widows and retired executives, leave it to their lawyers, stockbrokers, or such organizations as the very minor minority organization (immorally misrepresenting themselves as the "Moral Majority") to make their voting decisions.

Regarding such doing-the-thinking-for-others organizations as the "Moral Majority," we have Yale University President A. Bartlett Giamatti's statement in August 1981, as quoted from a United Press International dispatch:

He said that the Moral Majority and other "conservative" groups are "shredding the spiritual fabric of our society" and are "intent on destroying diversity of opinion. . . . They threaten through political pressure or public denunciation whoever dares to disagree with their authoritarian positions. . . . They would sweep before them anyone who holds a different opinion." Giamatti further criticized the "peddlers of coercion" for pressing uncompromising attitudes that are "dangerous, malicious nonsense" and as being advocates of "polyester mysticism" with a goal to "divide in the name of patriotism. . . . They have licensed a new meanness of spirit in our land, a resurgent bigotry that manifests itself in racist and discriminatory postures."

The supranational-giant-controlled Madison Avenue advertising billions at once replied through the editorial pages of its U.S. newspapers which undertook to vitiate the statement of Yale University's president. It is not likely that the third-oldest and second-richest university in America has a board of trustees and faculty that would select as its president an irresponsible "crackpot," which many media editorials asserted Giamatti to be.

Because many of the invisible supranational corporation's manufacturing facilities are presently located on U.S.A. geography, and the invisible giant's armaments objectives require ten years lead-time, the colossus has now of necessity placed its initial armaments orders in the U.S.A.-situated factories. Capitalism's supranational corporate colossus also finds it most convenient and invisibly expedient to continue doing its business under the name "United States" which is easy for it to do effectively: first quite simply by hoisting the American flag in front of all its factories, and second by pulling the vital strings of the finance-shackled and lobby-locked congressional puppets to make them pass the requisite legislation. Thus, the colossus now (early 1982) has in production the necessary first-things-first of its ultimate ten-year, multi-trillion-dollar procurement, world-published as being "U.S.A. national defense activity."

Throughout its first 127 pre–World War I years, the U.S. government often had no national debt. World War I left the U.S.A. with a national indebtedness of $33 billion. The U.S.A. banking system went truly bankrupt in 1929, but the New Deal's 1933 Bank Moratorium postponed recognition of that fact. Since then the moment of acknowledgment that the U.S. government itself is financially bankrupt has been postponed first by further- and further-ahead postponements of the payoff dates for U.S. notes and bonds and by successive votes of the U.S. Congress to increase the national debt limit. By "money accounting" (in contradistinction to real-wealth accounting), the U.S.A. is now realistically bankrupt. Since Nixon became president, the U.S.A. has been unable to pay even the interest on its national debt, let alone reduce the principal. Before Nixon, Congress assumed tax underwriting of ever greater interest-bearing on ever more postponed and greater national debt limits. For all the Nixon years and all the years of his successors the president has had annually to file a negative budget, meaning the U.S. cannot even pretend to be able to pay the interest on its indebtedness.

The supranational-money-colossus banking experts therefore have ordered its U.S. 1981 president and its congressional puppets to legislate stringent budgetary measures without reducing Grunch's annual earnings, to be accomplished by curtailing old-age medical aid and eliminating schoolchildren's lunches—eyewash measures permitting the U.S. government to "technically" assume that by 1983 it would be balancing its budget as a first requirement of "sound banking practice," warranting the banks loaning their funds to accommodate the most profitable known business, that of armaments. However, by 1982 it became evident that the administration was not going to be able to do so, even within roughly a $100-billion deficiency. Balancing the budget would, in fact, be only a "creditability" cosmetic in view of supranational-controlled international banking's superabundant monetary capability to pay for the ultimate $6-trillion CIA's (Capitalism's Invisible Army's) rearmament goal. If the supranationals are unable to accomplish a U.S. budget balancing, then a consortium of the supranational, only-logos-identified entities will use its own credit card banking system and comfortably charge off to insurance its $6-trillion armaments-acquisition program. But they don't want to play the game in that profitless manner. They want to do it through U.S. defense orders, which can pay handsome dividends throughout the process, making possible not only the continuance of all the corporation executives but the refunding of their capital expenditures.

Six trillion dollars! Let's try to sense the magnitude of that. Six trillion happens also to be the number of dollars the U.S.A. and the U.S.S.R. have already jointly spent on armaments since the end of World War II, when the United Nations was established. All electromagnetic radiation travels at 186,000 miles per second—that is, a little over seven times around the Earth's equator in one second, over to the Moon and back in four seconds, or over to the Sun in eight minutes. Six and one-half trillion is the number of miles radiation reaches out radially in one year, traveling at 186,000 miles per second. That is the magnitude of the number of dollars the supranational corporations are now intent on spending, and spending exclusively on killingry, at the same time that old people are deprived of their security and children go lunchless.

With control of the "Free World" media, the colossus hopes to postpone world realization of the fact of U.S. bankruptcy for ten years. Those ten years constitute the time needed to complete their program of rearmament procurement through U.S. government machinery.

Having attained its present military superiority, the U.S.S.R. is not going to give the supranationals those ten years to build their new offenses. The U.S.S.R. sought its superiority in order to force disarmament, so that it could turn its vast industrial productivity to the benefit of its own people to prove that socialism can demonstrate a higher standard of living than can capitalism, and knowing that capitalism is intent that the U.S.S.R. shall never have the opportunity. Knowing that capitalism is so intent, the U.S.S.R. is probably going to try to force a conventional warfaring showdown long before that critical ten-years-to-regain-parity opportunity is afforded to capitalism. Capitalism's recourse is to have its leaders move into the Rocky Mountain atomic-bomb-proof apartments, then to "push the button which ends it all for everyone."*

---

*If you are upset by that, don't forget that *Critical Path* does offer a happy, but only by-the-skin-of-our-teeth to be attained "out" from the dilemma.

Chapter 5

Paper-into-Gold Alchemists

The supranationals have now completely forsaken their leadership in the once-upon-a-long-ago-time, pro-humanity, industrial mass-production, gained exclusively through individual inventive ingenuity, integrity, and local community pride in producing only the best possible products, as does Japan today. Such was the leadership of Henry Ford, Sr., who was inspired to mass-produce no-frills, reliable motor vehicles for the lowest possible prices primarily to help the farmer get to market. That his activity involved large amounts of money was only incidental. It was obvious to Ford that a prudent amount of earnings must be set aside to buy ever-improving equipment. Also, he determined that a safety-factor surplus be set aside against poor economic days. Ford's enterprise was never "to make money." At enormous expense he bought back all the shares in his Ford Motor Company from his original backers, whom he found were primarily interested in making money. Henry, Sr., fought J. P. Morgan for many years as to which it should be, "make sense or make money," which are mutually exclusive.

Ford's son and grandson failed to understand old Henry's inspirational philosophy of real-wealth producing and learned to play only the game of moneymaking with the money they inherited.

It is the strategic prerogative of the invisible corporate giant to unilaterally and arbitrarily alter the scoring values in the economic game of "earning a living" vs. either "winning a living" or "tricking a living." The corporate colossus alters the scoring values by increasing prices to ensure that the industrial game will always be won only by the "richest." It is reminiscent of the following incident of my boyhood.

Amongst the neighborhood boys of my childhood was one whose family was very rich and had bought him a set of baseball bats, balls, gloves, mitts, and other equipment such as catcher's masks, base cushions, pneumatic catcher's belly protector, home plate—that is, the equipment for an entire team. Finding a suitable neighborhood field, he would attract a crowd of us. Eager to play, resplendent in his baseball uniform, he would announce the rules of baseball as we were to play it if we wanted to use his equipment. When, however, his side or he himself began to lose or play poorly he would change the rules, making his side's just-scored runs worth five of the earlier runs. If his rule-switching became unacceptable to the rest of us, as it often did, he would pick up all the playing gear, put it in his pony cart, and drive away.

Here is another example. In between the "halves" of an end-of-season championship professional football game, the home team's super-rich and powerful owners, finding their team three touchdowns behind the visitors, convene an effective quorum of the league's team-owners present.

Using various economic wrist-twisting tactics, they persuade a "carrying" majority of the owners to vote that henceforth until further notice touchdowns are to be scored thirty points each, including the one touchdown just made by the home team. As the final scoreboard reads 30–21 in favor of the home team, it is easy to imagine not only the consternation of the public over such unfair tactics but the public's outright rejection of such on the spot, unilateral value-changing. This, however, is all there is to the phenomenon known as unilateral price-advancing, which is solely responsible for "inflation."

The president of U.S. Steel Corporation says, "The price of U.S. steel is now advanced ten dollars per ton." The president of the United States says, "Mr. U.S. Steel, you can't do that." The president of U.S. Steel replies, "Not only can I, but I already have—and that's that, Mr. President. Over and out!" Clonk.

Individual corporations within the economic power structure bring about inflation through unilateral price-advancing, about the "reasonableness" of which unilateral score-changing the corporate greats use their media ownership to educate the public. They claim their price increase has been brought about by aggravating cost increases within their particular industry as well as by foreign competition, ergo by evolutionary events beyond the price-increasing management's control. To avoid anti-trust action, the price increases are done by the great industrial corporations independently, one by one.

•　　•　　•

In order to understand the economic events transpiring in 1981, we must comprehend the rudiments of the financial securities world. The "abstract-being" corporation operating within the United States receives legal authority from state and city governments to sell common shares in "outright risk" ventures, to which common shareholders they promise only proportional sharing of cash profits from their economically successful operation. But corporations also receive legal authority to sell "preferred stocks," to the owners of which the corporation becomes legally obliged to pay a fixed annual rate of interest before distributing any of the profit to those corporations' common stockholders. If, however, the corporation has no annual profit, it has no obligation to pay interest to preferred stockholders. "Preferred" means first to be "paid off" at a fixed annual percentum rate out of net earnings, after which fixed percentum pay-off the common shareholders divide as much of the earned profit as the directors of the corporation feel can be distributed while also taking care of new-development expenses as well as safety-factor reserve fund set-asides.

If the corporate enterprise is to be liquidated, the preferred stockholders are to be paid off before the common stockholders. If the corporation fails, the preferred stockholder share in whatever equity may be left, but not so the common shareholders. If the corporation fails and no assets remain, the preferred stockholders have no further legal recourse, i.e., claiming rights.

In addition to the common and preferred stocks, both of which are outright risk-venture shares, corporations can raise initial, working, or expansion capital by issuing bonds, provided the corporation owns free and clear real wealth, that is, real estate or buildings and easily salable general machinery, such as machine tools, lathes, drill presses, etc., with which machinists produce special-purpose mass-production tools. The word *real*—of "real" wealth, of "reality"—developed from the Spanish word for "royal" or "royalty." The king or queen personally acknowledged truth. *Real* is what the socioeconomic power structure decrees it is. *Real* is, like the abstract-legal-entity corporation, a legally accepted fiction—real estate equals royal estate.

Bonds are called securities, and their value is predicated upon the degree of probability that upon public liquidation of a company, the real estate, buildings, and machinery can be sold at the bonds' face value or more. While corporate bonds have priority over preferred and common stocks when corporations are liquidated, fail, or are sold, the bonds do not have priority of claim over those to whom the corporation is already indebted for goods, services, etc., nor do the bonds have priority over federal, state, or municipal taxes. Bond-owners have no share in the earnings of a corporation but do have fixed-interest priority over preferred stocks.

It is relevant to our understanding of securities to note that bonds other than those of tax-guaranteed federal, state, county, and municipal bonds have often proven to be less than secure. For instance, approximately all the railroad-company bonds of the U.S.A. were defaulted, i.e., became 100-percent unredeemable, in the economic depression of 1929 to 1933 and thenceforth.

Within the United States of North America, the right to incorporate is legally obtainable only from state governments and not from federal nor city, town, or county governments. If, however, corporation promoters wish to sell shares worth more than $300,000 for capital expenditures before earnings, then the federal government's Securities and Exchange Commission must also give its approval.

Some states in the United States grant incorporation rights that are more workingly satisfactory to a given type of enterprise than the privileges granted to the incorporators by other states. Some state's incorporation privileges are extended to cover the corporation's operations in other states. Many states grant incorporation privileges only within their own borders.

In addition to the securities issued by corporations or by nonincorporated companies or partnerships, bonds are issued by towns, counties, cities, states, the federal government, and multi-state "authorities." Federal, state, county, and city (municipal) bonds have face declared maturity dates and clearly scheduled tax-collection commitments to cover both the yearly interest rates and the maturity repayments to the bond-holder. All such government bonds have prior right to funds produced by taxes.

In the world of securities—properties—represented by written documents that are legally recognized forms of negotiable monetary investments, there are also insurance policies, and in particular life-insurance policies.

Known in world history's earliest records as "bottomry," what we now speak of as insurance dates back to its practice in 4000 B.C. Babylon as the underwriting of merchant-ship voyages. This practice made it possible for wealthy individuals who were not shareholders in the original shipbuilding enterprise to participate at their own risk in the merchant-ship venturing, which when successful was fabulously so.

Historically, life insurance is a very recent form of the already wealthy humans' and their corporations' capital gambling, and does not begin until the industrial era of the late nineteenth and early twentieth century, when technology began to render humans' immediate physical environment more propitious for the protecting, support, and omni-accommodation of increasing numbers of human individuals, despite remote devastations occurring in the world's wilderness areas.

The technology of industrialization was first taken advantage of by the military specifically intent on how to more accurately kill more and more people at ever great distances in ever shorter periods of time. As a consequence, the technologically advanced nineteenth-century weaponry witnessed greatly increased war-wrought devastation, leaving more and more wounded to die on the battlefields. The medical world being as yet inexperienced and ignorant, the U.S.A. Civil War saw the highest ratio of deaths per battlefield-committed soldiers in all history. Due to medical ignorance, the U.S. Civil War also saw more wounded left unattended to die on the battlefield than in all other wars of known history.

When the U.S.A. entered World War I in 1917, it was asked by the British to replace all the line-of-supply ships the British had lost to German submarines and simultaneously to bring the U.S. Navy to parity with that of the British while also training, arming, transporting, and Navy-escorting one million soldiers across the German-submarine-infested Atlantic to the battlefields of France. When the numbers of U.S.A. troops killed and wounded in World War I battling reached unprecedented numbers, the U.S. Congress was confronted with the enormous cost of training, arming, and transatlantic-replacing of their killed and wounded troops in France. The U.S. Congress was then informed of an alternative solution to the replacement problem.

The U.S. medical scientists informed the U.S. Congress of the potential ability to save and repair the wounded U.S. soldiers in France, provided enough money was appropriated for a known-to-be-possible vast advancement in medical science: drugs, equipment, and practice. The cost of this capital investment in medical science, though historically unprecedented, was far less than the cost of entirely new troop replacements from the U.S. Convinced of these facts, the U.S. Congress appropriated the funds for the medical-science solution of its problem.

It worked. When the war was over and the saving and rehabilitation of vast numbers of veterans was realized, the new-era medical establishment was not disbanded. Enthusiastically supported by citizens in general, scientific medicine refocused its attention on the U.S.A. home front. One after another, the immediately fatal and "incurable" diseases, lethal conditions of yesterday, came swiftly under complete control. Medical information regarding further curing and effective anticipatory avoidances was enormously expanded in the late 1920s. It was discovered in the late 1920s that the area of highest mortality was the period of childbirth and its first ensuing four years. This brought successful coping with these initial years' fatalities into general effectiveness in the 1930s. The seeming population explosion after World War II was due in fact not to a postwar increase in the birthrate, whose small rise in the U.S.A. lasted for only two years, but to the coming of visible age of those who used to die but did not now or hereafter die in the womb or at birth or within their first four years in the 1930s, as had those conceived or born before the 1930s, together with the subsequent escapees of the pre-1930s childhood mortalities.

There is no scientifically accredited manifest of an alteration in the extreme-limit magnitude of human life-span throughout all recorded history. There are many claims and assertions of great longevity, but 113 years is the greatest span positively known to exist. One hundred and thirteen years is the terminal norm of human life. How many attain their potential longevity is another matter. Few do. However, the number of those who do is now increasing. Quite clearly, the ability to substitute plastic bones and mechanical hearts, etc., suggests that we may have entirely inanimate physical human mechanisms, proving what we have contended from the beginning to be true, i.e., that whatever it may be, life is not the physical equipment that it employs any more than is the telephone or any other of the integral or detached tools that life employs.

Despite many dramatically negative side-effects and feedbacks, twentieth-century technology has rendered the everyday physical environments in the U.S.A. and elsewhere ever more favorable to human life. This has greatly increased the realized life-span as well as the physique of Americans and others living under the same technological improvements of the environment—in the years between World War I and World War II, the average height of American males increased four inches.

A popular 1900 song said, "A dollar a day is very good pay when you work on the boulevard." Two and a half dollars per fourteen-hour day was the pay for my own first pre–World War I job in New York City with one of the great meat-packing companies, $15 a week ($880 a year), and I was able to live happily but frugally on it. I was married in 1917, and as an ensign U.S.N. in the U.S. Navy's European war zone duty I was receiving $1,800 a year.

The younger you are, the more attractively low are the annual rates of payment for maintaining each thousand dollars worth of a life-insurance policy. In 1917 I figured that if, when I died, my life insurance policy would give my wife a capital sum of $50,000, she could deposit that sum in a savings bank. She would then receive a 4 percent annual interest payment of $2,000 which would be $200 more per year than had been our U.S. Navy ensign's pay. She would have only one of us two to look out for, so that she should be able to save and add to the principal deposited in the savings bank. According to their advertising, the savings bank would continually "compound" the interest on the deposited amount. That was the economic picture presented to me in 1917 by the so-called honest businesses and the deeply respected banking world. Prudential Life Insurance Company depicted itself as "Secure as the Rock of Gibraltar." So I bought $50,000 worth of life policies. For my World War I zone activity of eighteen months duration I was given a bonus of a fully paid $5,000 of life insurance payable to my heirs at my death. The life-insurance company's sales talk failed to anticipate the great deflation of the individuals' dollars, brought about by moneymaking business' greed increasing the prices of products and services whose physical costs were technologically ever decreasing.

A third of a century later, my annual cost of living at a level which permitted me to travel to give lectures and to design and physically produce improved mass-production prototypes of livingry artifacts reached my 1917 assumed total "live-on-its-interest" capital equity of $50,000 being completely spent each year. I sold my life-insurance policies to the life-insurance companies for their then accrued cash value of $22,000. (This bought me the time to develop the geodesic domes.)

Today, as yet living in the same manner while doing the work essential to following through on my life commitment to solve socioeconomic problems exclusively by invention and development of artifacts instead of by politics, my annual cost of operation has reached $300,000 for servicing my commitment to problem-solving only by artifacts with no savings accruing whatsoever. Because I am not a "business" or a corporation my gross intake is rated as personal income and I am therefore in the 50-percent tax bracket, which precludes my saving any of that income, should I wish to, which I fortunately do not; though I am convinced that I am getting more for humanity with the dollars I spend than is any form of tax-supported government expenditure.

Each of the three Dymaxion three-wheel, front-drive,

rear-steered cars, built by me in the depths of the 1930s depression, cost $28,000, i.e., $84,000 in all. Today they would cost twenty times as much for the same vehicles, built of the same materials.

Boastfully "ethical" big business and banking have reduced my "necessitous and desirable" acquiring capability by 95 percent. That is, they have priced and otherwise manipulated the money game in such a way that $95 out of every $100 I have earned has been taken away.

Life-insurance companies bet that humanity is going to live longer than it expects. Those who buy life insurance are betting that they are going to live shorter than the insurance company thinks they are going to live.

The life-insurance companies invented the term "expectancy" of probably-to-be-realized longevity of specific classes of humans existing under various environmental conditions in contradistinction to the (now workingly assumed) unvarying maximum limit of 113 years for all people.

The overall history of humanity's attained longevity is one in respect to which the average individual's attained life-span has progressively increased from a "probable average expectancy" of nineteen years of age for the people of 5000-B.C. Egypt (at sixteen years, Alexander the Great was the world's leading general) to seventy years in A.D. 1980 U.S.A. Though manifesting many setbacks, the overall evolutionarily inexorable increase in longevity of those living within the technologically advancing environment is a product of humanity always gaining more experience and learning ever more from its errors of conceptioning, and acting on how to cope more effectively and intelligently with the progression of natural exigencies. Humans gain despite enormous setbacks by industrial environment-polluting.

The insurance companies collect all possible vital statistics governing the probabilities of individual human survival under all the known controlling conditions. The insurance companies have large staffs of statisticians known as actuaries, who, aided today by computers, are able ever more effectively to render the insurance corporations' stockholders bets to be approximately "sure things." In investing the insureds' premiums today, they find IBM, DuPont, and other such stocks to be "sure things," and tax-secured U.S. government, state, county, and municipal bonds the most risky of investments, for reasons we will later examine.

During the first three-quarters of the twentieth century, average life-expectancy in the most favorable environment countries has almost doubled. When I was born in 1895, life expectancy for a male born in New England was forty-two years of age. On July 12, 1933, I passed my "expectancy day." On July 12, 1979, I completed my "second lifetime" or double expectancy. I am now in the third year of my "third-expectancy lifetime" and am very healthy, with a new stainless-steel hip I have just acquired.

My two completed lifetimes and my third of a third lifetime have found the great majority of "savvy," well-to-do individuals I have met convinced that there exists an inherent inadequacy of life-support on planet Earth, and therefore that their own successful survival as well as that of those whom they cherish depends upon their cleverly learning more and more about how to be legally selfish and thereby to accomplish personal economic advantage by anticipatorily depriving others in directly undetectable ways. These ways are legally and socially accepted practices of deceiving and cheating the public—e.g., by altering the scoring system and the official "game rules" of the accrued monetary equities of other humans through zoning laws, "etc." $\times$ 10^{10} ways.

Price manipulation is most often defended as being governed by supply-and-demand variables, i.e., by what the "traffic will bear" and not by time-energy costing, which science finds to be exclusively operative throughout all the constant energy-transformings and interchangings of Universe. This legal deprivation of other humans to one's own personal advantage is most simply accomplished through increasing rents or prices of a cup of coffee or a second-rate cigar (both of which have escalated during my lifetime from five cents to fifty cents). The powerful social precedent for price-advancing has been initiated by non-perishable mine and oil-well owners and those other prime industrial producers, the physical cost of whose products, measured in ergs of energy per hour and pounds of coal or petroleum per pound of manufactured goods, has steadily decreased.

The acceleration in technological enhancement of the living environment apparently accounts for marked life-expectancy increases in post–World War I Canada, U.S.A., Sweden, Australia, United Kingdom, New Zealand, Netherlands, Japan. In these countries in 1968 the value of life-insurance policies in force exceeded their respective gross national incomes. In the U.S.A. in 1970 there were 355 million life-insurance policyholders. This being more policies than citizens meant that many citizens hold several policies. In 1970, the assets of 1790 U.S. life-insurance companies totaled 208 billion dollars.

Throughout the history of the United States of North America, until 1952, the federal, state, and municipal bonds guaranteed by the tax-collecting capabilities of those authorities, fortified by their ability to seize the property of tax evaders, were generally rated the most secure of "securities" to protect the U.S.A. population.

Until 1952 the states and federal government required that all life-insurance companies, all savings banks, and all trust funds invest only in notes and bonds of the ever-suitably-increasable, tax-supported federal, state, municipal, and multi-state authorities, whose bonds were officially qualified by federally supervised authorities as being "legal for trust funds." Included in the trust-funds category were all the bank- or lawyer-administered wills and trust funds of all kinds, all employment or executive retirement funds of corporations, labor unions, or cooperatives, all savings banks and life insurance companies.

In the post-1933 pullout from the "absolute economic crash" years of 1929 to 1933, the U.S. Congress approximately unanimously enacted a number of measures essential for coping with the economic errors clearly manifest in the post-mortem studies of the Great Crash. Essential to the correction of those errors was the establishment of absolutely interconstant price, wage, rent, and interest-rate controls. Profits were not only permitted but welcomed. They had, however, to be inventively attained by producing more, better goods, more satisfactory services, etc., for the same or lesser amounts of energy, time, and materials invested.

Thorough review of those 1930–42 economic climb-out years' events and their utter undoing in 1952 is clearly related in the chapter "Legally Piggily," in my book *Critical Path.*

In 1952, the twenty-year moribund G.O.P. regained control of the political initiative and immediately eliminated all price, rent, interest, and wage controls. It also passed laws allowing insurance companies, savings banks, retirement funds, and all trust funds in general to invest in common stocks, preferred stocks, and corporate bonds. This also brought about the formation of a myriad of investment-trust funds. It did not, however, allow the U.S. government or any of the states, counties, or municipalities to invest the enormous Social Security automatically-tax-collected individual worker's funds or any other of its future-committed economic-responsibility funds to be invested in any other than its own U.S.A. government, state, county, or municipality dollar-savings-accounts, whose purchasing equity was being ever depreciated by inflation.

With wholesale industrial prices freed to escalate (though protested against by the U.S. Eisenhower-through-Carter presidents), retail prices, wages, and corporate share prices as traded on the stock markets swiftly escalated in response. So-called inflation was inevitable. What the colossus' media call "inflation" is of course deflation of humanity's buying power. Inflation does not increase the true values nor produce more or better goods.

All talk of Federal Reserve rediscount-rate-increases acting as an inflation retardent is in fact only "rationalization" camouflage for escalating bank usury rates. Though a Federal Reserve interest-rate increase may produce a momentary deflection in the "rate of inflation curve," it never does any more than that. There is no evidence whatsoever that Federal Reserve rediscount-rate-increases successfully arrest inflation. This being so, the continuance of such interest-rate increases ostensibly to combat inflation as actuated by the private-enterprise-controlled and deceivingly misnamed "Federal" Reserve Bank system must in historical retrospect be identified as a fraudulent means for increasing the profits of the banking system.

When Nixon cut loose the U.S. dollar from its U.S. government 1933-fixed ($32-an-ounce) relationship to gold, the U.S. citizen's dollar equity declined precipitously.

The last third of a century of overall stock-market *price-advances,* in direct correspondence to the role of inflation in business pricings in general, has produced ever widening but false gaps between people's government fund equity values and the portfolio of world-around stock-market values of the securities of the private-enterprise system.

Had the U.S. government in 1952 been allowed—by the U.S. government itself, which did the granting—to invest their Social Security and other appropriated funds in *Fortune*'s leading one hundred corporations' common shares, as were all the fund managers of all the within-the-U.S.A. trust funds, life-insurance companies, and investment trusts with their funds, the 1982 value of the U.S. government-tax-collected Social Security and other funds would permit the U.S.A. to extend all manner of social benefits to all its people.

Hopefully, to prevent unilateral price manipulation the original value of the U.S. government tax-collected Social Security funds was price-locked into gold at $32 an ounce. This was done when approximately all the world's monetary gold bullion was stored in the U.S. Treasury's Fort Knox, Kentucky Hills vaults. It was then that the U.S. government contracted with its working citizens to provide each with post-retirement Social Security, to be purchased by them—as with all life insurance—by payments deducted from their wages and salaries and originally matched by equal sums paid for by the employers, all paid to the government in the form of taxes. The first Republican presidency after the New Deal allowed the employers to write off their share of this by allowing them to deduct it as operating expense before calculating annual taxes.

The January 1982 value of the U.S. government's up-to-the-minute paid-in Social Security and other tax-collected, government-committed funds are now stashed away only in ink figures in the U.S. Treasury Department books, completely bereft of that fund's $32-per-ounce linkage with gold, as arbitrarily severed by the Republican president of the United States without authority from the U.S.

Congress and only on the advice of media-unidentified, non-government "others" with whom he met in a secret two-day conference on Minot Island in Gilkey's Harbor, Islesboro (a Penobscot Bay, Maine, island), after most of the gold had been withdrawn from the Kentucky Hills vaults to meet negative international balances of trade brought about by oil imports.

If, since their inception a half-century ago, the tax deductions for Social Security had been invested in leading common stocks by a consortium of Merrill Lynch–grade investment houses, the majority of U.S. senior citizens would now be multimillionaires.

The advertising industry, the brokers, the banks, the conglomerates, the shipping business, and labor unions have all insinuated their money-making or wage-equity equalization into the complex price-increasing. All the while, the actual basic costs of nature's cattle's calves' hide-making from the photosynthetically Sun-impounded hydrocarbon-rich vegetation energy has not increased one iota. Inventors have greatly reduced the manufacturing costs accomplished only by machinery whose installation costs require large capital sums.

What we have experienced are ever larger sums of other's money being commanded by *the money-makers*— all at the expense of the individual human beings who do not command such sizable funds, discouraging their enterprise and initiative.

When in 1920 I bought my $50,000 life-insurance policies, the best U.S. Navy officers' shoes cost $4 a pair. Civilians had to pay $5 for first-class leather shoes. Fancy, made-to-order Cordovans cost $7 a pair. In the sixty-year interim, the cows have not improved the quality of their hides; automated shoe-manufacturing processes have not improved the shoes. In fact, the BTU's (British Thermal Units) per each kilowatt of the electricity required to operate the shoe-production machinery has continually decreased; but money-making business has gotten the price increased to $50 a pair. To do so, after each of their arbitrary price-increasings solely for greater profit they have had to share a minimum of their profit with the labor unions else the workers could not buy the shoes. If you are not a labor-union-backed wage-earner you simply pay more and more for the same shoes, which again means that those manipulating large amounts of money are robbing you.

The stock markets reflect these price increases. Since the shoes are no better and often worse, the net of it all is that the same shoes require ten times more of the individual's paycheck dollars, so I am being robbed by the system and my government is unable to protect me; in fact, the present administration of my would-be people's government is being conducted by those who have been robbing me. They have become utterly callous in their viewpoint of rationalized selfishness based on the economic assumption of "not enough for both you and me" wherefore they look out only for themselves. These are the people who are even now maneuvering toward an atomic-war exchange in which all the unemployed and poor in general are the only ones sure to be eliminated.

The Reagan government's reduction of taxes for the rich money-makers which might otherwise have been applied to saving the nation's human security system, while concurrently also trying to increase the annual amount of armaments appropriations to a $150 billion level, shows how absolutely ruthlessly the directors of the supranational colossus are determined to acquire their own super-arms protection against communism while realistically making hundreds of billions of dollars a year manufacturing their armaments while also ruining the lives (slow murdering) of many millions of U.S. citizen workers to whom the U.S. government has sold the insurance of old-age and health-security benefits. Nixon's cutting loose of the price anchorage betrayed all non-wealthy U.S. citizens.

The Reagan government has reinstated the military draft to raise troops to protect capitalism against communism, while refusing funds to bury the bodies of its appropriations-abandoned deceased veterans, which again makes it clear that what was once the Republican Party of Lincoln is now the party of unmitigated selfishness of big money.

We have arrived at an overall economic condition where the seemingly inalienable rights of U.S. citizenship can be enjoyed only if one is the owner of an amount of stocks whose dividends exceed $50,000 annually per individual at the present level of dollar depreciation. The very word "inflation" is a deliberately deceptive term adopted to exploit the easily misguided human reflex conditioning. It was spontaneously chosen to obscure the fact that all nonholders of corporate shares are being legally robbed.

The U.S. government Treasury, in its almost daily task of arranging for loans from syndicates of banks and brokerage houses for funding and refunding its ever-more-frequently-coming-due-for-payment short- and long-term government obligations, finds it ever more expensive to market even its short-term promissory notes and longer-period bonds. Its long-term bonds sell at prices which, with interimly accruing interest, have ultimate monetary yields—between the price at which they were purchased in the financial market and their face-value maturity payoff—in the neighborhood of 18 percent, showing clearly that the financial-market world is now assuming that the U.S. government will soon reach a crisis point beyond which it will no longer be able to pay off either its short- or long-term obligations. All the big brokerage houses and banks that join in syndication to handle the sale of the government-refunding-note sales make so much profit in doing so that they will keep on risking their joint underwriting until they see the moment of formally acknowledged bankruptcy of the nation to be less than a year away. When the score reads one trillion and a half, with a debt increasing regularly at over 100 billion a year, somebody is going to see that the emperor has no clothes and when they shout out to that effect, there will be a swift world acknowledgment of the fact. Then they will try "Title 13," which allows the bankrupt organization to keep on operating and thus terminate any and all risks of private enterprise—for which risking on behalf of human gains it has been given the many considerations such as corporate privileges.

In the September 1, 1981, sales of U.S. Treasury ninety-day notes, which were almost certainly redeemable before any U.S. government bankruptcy payment-defaults, produced yields that were momentarily higher than the earnings on many high-grade corporate shares. Until the summer of 1981 these short-term, high-yield government notes had been sold only in minimum lots, the prices of which have been beyond the everyday buying scope of individual citizens but at negligible prices to the supranationals or any of their world banks. It is now far more risky to buy U.S. government securities than it is to buy shares in what once was risk capital's economic venture corporation shares. There is nothing in the words or spirit of the U.S. Declaration of Independence or U.S. Constitution which states or suggests the U.S.A. is committed exclusively to the success of the rich. The U.S.A. we have known is now bankrupt and extinct.

The swift coming-apart of our national economic system to make way for evolution's inexorably developing integration of a world socioeconomic system is manifest in the full-page advertisement in the Sunday *New York Times* (August 9, 1981) by one of the most conservative of the New York savings banks: "18% Interest Starting Now. Up to $2,000 Tax-Free Interest Starting Oct. 1." For this same interest rate, Dime Savings Bank required an investment of $5,000. Twenty similarly unprecedented savings-bank ads appeared in the same financial section. Before 1980, U.S. savings banks had paid in the range of 4-percent interest. The following week, that of August 16, the same plethora of bank advertisements appeared, with the percentage increased to 25 percent. By the week of August 23, the advertisements had advanced their offer to 40 percent interest. This was, of course, the correctly calculated percentage of gain, but misleadingly employed. The percentage represented "yield to maturity" in respect to the low August purchasing price of U.S. Treasury notes maturing on October 1, 1981, whose depressed market value indicated not only the imminence of U.S.A. bankruptcy, but the expectation that this particular set of U.S.A. "refunding notes" at ever higher interest rates and on ever shorter maturity terms would in all probability be redeemed before common recognition of the outright bankruptcy. The advertising was deliberate and mischieviously misleading in that it implied a continuing 40-percent interest rate if you deposited several thousand dollars— whereas the interest advertised was to accrue only until October 1, 1981. The advertisements failed to clearly communicate that the interest rate after October 1 would be a customary savings-account rate. The advertisements were hoaxes to acquire savings deposits, which the bank could loan out at much higher rates of interest.

At this point in our review of how economic and other social conditions have evolved to such a threateningly devastating future outlook for many if not all humans, we wish to recall that inexorable cosmic evolution is intent on integrating all humanity in one global government and, therefore, on eliminating all of planet Earth's nations and on doing so in a hurry. The most difficult of all the world's sovereignties to eliminate is clearly that of the U.S.A. We recall having forecast this termination of the U.S.A. at least fifteen years ago. The 150 nations are 150 clots in the economic bloodstream of our planet. The headlong rush into the atomic holocaust is in fact a far more threatening development than the natural economic demise of the U.S.A., which in fact may be viewed as simply a self-removing-planetary-economics-blood-clot event.

Chapter 6

Can't Fool
Cosmic Computer

It must be remembered that, as clearly elucidated in *Critical Path* and *Operating Manual for Spaceship Earth,* * money is not wealth and that wealth is the organized technological capability to protect, nurture, educate, and accommodate the forward days of humans, whereas money is only a medium of exchange and a cash accounting system. Money has become completely monopolized by the supranational-corporation colossi, which inher-

---

*The July 31, 1981, issue of *Publisher's Weekly* reported on the first U.S. publishers' trade fair in the People's Republic of China. Over six hundred publishers and eight thousand titles were represented, with several million English-reading Chinese attending the exhibit in six cities. Elaine Frumer, a member of the U.S. delegation reporting in an accompanying article entitled "What the Chinese Looked At," listed *Operating Manual for Spaceship Earth* first on the list she compiled of the ten books the Chinese showed the most interest in.

ently as legal abstractions ignore the problem of how to protect and nurture human lives.

In the very largest way of looking at planet Earth's socioeconomic-evolution events, we must observe that humans are designed with legs and not roots. Yesterday, humanity developed temporary roots as it cultivated its life-support food root-grown on the land. The metals made possible metal canning of food and mobilization of machinery. Today, all of human existence depends on the swift, world-around intercommunication system operating at 186,000 miles per second. We have transformed reality from Newton's "at rest" norm to an Einstein's 186,000-miles-per-second norm. Socioeconomically we have synchronized with the omni-intertransformative kinetics of the entire Universe.

Planetary economics has now shifted from a physical-land-and-metals capitalism to a strictly metaphysical, omniplanetary, omnicosmic-wealth know-how capitalism. The once noble and essential but now obsolete nations belonged to the rooted socioeconomic land-capitalism era of humanity. In reality, humanity is now uprooted kinetically and occupying the whole planet. Capitalism is dumping its immobile real estate and depending on science to synchronize its affairs with the invisible realities, misassuming, however, that science knows what it is all about. To successfully dump all its real estate, capitalism has all but ceased "renting" and through enforced selling of "cooperatives" and "nothing else but condominiums" is forcing the citizenry into anchored exploitability, while it is always increasing the *corporate* deployability and mobile shift-about-ability around the world.

Since science and human inventiveness are continually learning how to produce ever more, and ever better performance, for each ounce of material, erg of energy, and second of labor and overhead time invested in any and all of industry's production functions, the real cosmic costs are always and only decreasing, and all price-increasing, as already noted, is corporate selfishness "gotten away with" by political-campaign obligations and by excruciatingly painful, behind-the-scenes corporate lobbyists' congressional bullying.

In 1940, two years before the U.S.A. entered World War II, the president of the Aluminum Company of America became interested in my use of corrugated aluminum in the Dymaxion Deployment Unit, which was the little, mass-producible, autonomous, one-room dwelling machine that I had developed for a group of Scottish leaders, intent on anticipating the wholesale bombing of England's industrial cities, who had proposed accommodation of the surely-to-be-displaced population in thousands of my Dymaxion Deployment Units to be installed on the Scottish moorlands. I was developing these deployment units at Butler Manufacturing Company in Kansas City. I was doing this by converting Butler's mass-production, twenty-foot-diameter, galvanized-steel grain bins into an autonomous, fifteen-hundred-dollar, well-insulated, fireproof, earthquake-proof, kerosene-ice-box-and-stove, Sears Roebuck–furnished, ready-to-move-into unit.

The Butler grain bins had been developed for the U.S. government "ever-normal-grainery" program. The Alcoa president hoped that I would switch from corrugated, galvanized iron grain bins to corrugated aluminum bins. He told me that the cost of aluminum would always decrease. He said that the main cost was that of the electrolytic refining of bauxite ore into aluminum. He said the cost of electricity for the process was always decreasing as we learned to produce ever more kilowatt hours per each BTU of fossil fuel expended. Aluminum, he said, is one of nature's most abundant elements. The wholesale price of aluminum in 1940, when Alcoa's president made that statement to me, was twelve cents per pound. Today it is selling for over a dollar a pound, not because the Alcoa president was wrong in what he said but because massively organized selfishness has dishonestly changed the scoring system.

It is evident that the degree of technical "advantage" now attained by world-around industrial production capability, if realistically appraised and articulated, now shows that all humanity has just reached a state of comprehensive technical advantage adequate to provide a billionaire's level of living on an indefinitely sustainable base for all of the over four billion human passengers now aboard Spaceship Earth (see *Critical Path*). The world's economic accounting system, if properly entered into the world's computers, will quickly indicate that comprehensive economic success for all humanity is now realizable within a Design Science Decade. All it takes is shifting from weaponry to livingry production.

History's unprecedentedly large and invisible supranational Grunch of Giants being too supra- and infra-visibly large to be sensitively comprehended, it is difficult to surmise and accredit that the almost omni-computerized giant may be evolution's agent of most effective establishment of a world-embracing socioeconomic system most logically suited for the mass-production and distribution of its products and services to economically successful humanity. It could well be that the total-world-involved, supra-national giant corporations' computer operations might, to their corporate directors' astonishment and to popular surprise, lead the Grunch into profitable discard of all that is not true, as for instance that anybody owns anything. Commonly acknowledged operational custodianship and popular reaccreditation of the integrated world-around technology management may supplant "ownership" with Hertz and telephone-renting.

The way that the giant can be successfully led into doing so is for a substantial majority of humanity, and eventually all humanity, realistically to comprehend the falsity of the greater part of the inventory of academic premises and axioms upon which the thus misconditioned reflexing of "educated" society is based. For instance, there is no God-validated deed to property of any kind whatsoever. There are no solids. There are no things—only systemic complexes of events interacting in pure principle. There is no up or down in Universe. There is no cubic structure. There are no straight lines. There is no one-, two-, or three-dimensionality. There is only four- or six-dimensionality, etc. As we eliminate that which isn't true, we inadvertently admit into reality that which is true. As world society divests itself of that which experimental evidence demonstrates to be untrue and embracingly enters into its computer the mathematical formulae of all that can be experimentally proven to be true, all the socially, selfishly malignant characteristics of the giant may vanish and the omni-pro-social-advantage-producing capabilities may prevail and flourish.

•   •   •

FIFTEENTH-TO-NINETEENTH-CENTURY
GIANTS' OATH
  Fee-fie-fo-fum
  I smell the blood of a Britishman
  Be he alive or be he dead
  I'll grind his bones
  To make my bread.
TWENTIETH-TO-TWENTY-FIRST-CENTURY
GIANTS' OATH
  We-Fort-five-hun
  Steal kudos and credit American

Be it live checking
Or savings "dead"
We knead their dough
For dividend bread.

● ● ●

Each of the giants of today's great Grunch is a quadril-lionfold more formidable than was Goliath. Each is entirely invisible, abstract, and completely ruthless—not because those who run the show are malevolent but because the giant is a non-human corporation, a many-centuries-old, royal-legal-advisor-invented institution. The giant is a so one-sidedly biased abstract legal invention that its exploitation by the power structures of thirty generations have made the XYZ corporations and companies seemingly as much a part of nature as the phases of the Moon and clouds of the sky. Corporations operate on an unnatural economic basis that makes a successful Las Vegas roulette bet a trifling success. If you bet your money on the fortunate corporation, your bet is paid and repaid to you quarterly, continually, ad infinitum, often more copiously each year. Assuming an investment of 100 shares at a cost of $2,750 in I.B.M. in 1914 when the company was first formed, it would have grown by June 1959 to 59,320 shares values at $19,308,900, plus $1,089,000 in cash and stock dividends and an additional $101,906 from rights and privileges.

As I have frequently recalled, the grossly mis- or under-informed, 95-percent-illiterate world society of 1900 misassumed the existence of a fundamental and dire inadequacy of life-support to be operative on our planet, wherefore it concluded, and the political-economic system as yet maintains, that it has to be only one or the other of the planet Earth's two great political ideologies which can survive, and that all the people governed by the loser must perish—"there is not enough for both."

The now-predominantly-literate world population of 1981 has developed an intuitive awareness of the illogicality and even madness of all political systems.

All the foregoing inadequate life-support misassuming by both political parties and all the major religious organizations, as earlier noted, has resulted in both sides having jointly spent six and a half trillion dollars in developing the present capability to destroy all humanity within one hour. Humanity at large is logically intuiting that the same sum spent in the direction of improving the lives of the presently deprived many might readily have brought about better results than race suicide.

The awareness of the emergence of a new world society has been only intuitive, because it is actualized only by a superficial knowledge of the overall integrated effects of an almost entirely nonsensorially contacted invisible reality of electronics, chemistry, metallurgy, atomics, and astrophysics. The epochal events of humans landing on the Moon, satellite-relayed "instant" around-the-world information, and an exclusively direct Sun-powered, Paris-to-London flight are altogether reorienting world-around humanity's intuitive thinking into the realization that we can now do so much with so little that we can indeed take care of everybody.

There is a deep urge on the part of vast numbers of world "youth"—irrespective of their years—to do something right now about their intuition, which develops an impatience and ever more volatile group psychology. There is therefore an urge toward open physical revolt even amongst some of those who do know there is a bloodless, design-science, revolutionary option to attain socioeconomic success for all. The hotheads want to yield to their impatience. To those who urge us to join forces in bloody revolution, I reply as follows.

Before humans could be designed to occupy it, planet Earth had to be designed. Before planet Earth could be designed, the solar system had to be designed. Before the solar system could be designed, galaxies had to be designed. Before special-case galaxies could be designed, special-case macro-micro Universe, all its atoms and molecules, gravity, and radiation had to be designed. Before *any* realizable designing was possible, it was cosmically necessary to discover and employ the full family of eternally coexistent and synergetically inter-augmentative, only-by-mathematical-equations expressible, intercovarying, generalized principles governing the generalized design of eternally regenerative scenario Universe. And before all recognition of the eternal generalized principles and their inherent design-science functions, it was further necessary to have:

1. The design of an eternally regenerative, radiationally expansive and gravitationally contractive, everywhere and everywhen complexedly intertransforming, nonsimultaneously episoded, scenario Universe.
2. The generalized design of galaxies of entropic matter-into-energy as radiation-exporting stars and generalized star systems of planets serving syntropically, as radiation-into-matter importing planets.
3. The design of planet Earth as the Sun-orbiting, biosphere-protected, and oxygen-atmosphere-equipped incubator of DNA-RNA design-controlled biological life and of that life's photosynthetic conversion of entropic radiation into syntropic, orderly hydrocarbon molecules and a vast variety of hydraulically compressioned, crystallinely tensioned, exquisitely structured biological species omni-inter-regenerating as an ecological omni-life and human-thinking support-system.

4. The eternal mathematics—numbering and structuring. The eternally extensive mathematical spectrums of frequencies, wave lengths, and harmonic intervals.

Only thereafter could those human beings progressively re-evolutionize exclusively by trial-and-error enlightenment, from their born-naked state of absolute ignorance, to discover their scientific-principle-apprehending minds and thereby (now for the first possible moment in history) to glimpse humanity's semi-divine functioning-potential as local-Universe critical-information-gatherers and local-Universe problem-solvers in support of the integrity of eternally regenerative Universe.

Only now for the first time can we human beings effectively revolutionize society in an adequate degree to fulfill what I have identified in *Critical Path* as being the number one objective of humanity's inclusion in comprehensive evolution.

The generalized principles have always—eternally—existed and have always been available for each special-case, revolutionary design-realization. The special-case design revolution has always had to precede the extra-special-case local social revolution, whether it be by inventing guns which overwhelm archery or inventing the wireless telegraph which transmits messages halfway around the world at six million miles per hour, making utterly obsolete the Pony Express and its concomitant "Wild West" socioeconomic behavior-patterning. The greatest evolution-producing revolutions are complex and take the longest to be realized.

•   •   •

What I hoped I had made clear in *Critical Path* is that the inherently half-century-long design science-revolution phase of attainment of universal economic success has been successfully completed and now needs only the *bloodless* socioeconomic reorientation instead of the *political revolution* to exercise humanity's option to "make it" for all.

I hoped that *Critical Path* made clear that fifty-three years ago I anticipated today's transition of humanity from 150 nations operating independently and remotely from one another to an omni-integrated world family, with all that socioeconomic transition's conditioned-reflex stressings and shockings.

I hoped that *Critical Path* made it clear that the world-data integration I initiated more than a half-century ago has kept growing into a comprehensive record of the invisible design-science revolution being achieved only by ever more performance, realized for ever less energy, weight, and time units invested per each increment of accomplished livingry functioning; and that more than a half-century ago I reduced my design-science, human-environment-augmenting structures and technologies to full-scale, physically working demonstrations of their advancement of technological advantage to economically accommodate an around-the-world pulsatively deploying-and-converging, kinetic society.

I hoped that *Critical Path* made it clear that if I or some other individual had not taken these anticipatory initiatives more than a half-century ago, the comprehensively integrated physical-gaining-for-all now to be immediately attained simply through inaugurating the mass-production phase of its already developed prototypings, then the design science revolution could not now be realizable—it would take another fifty years to do the critical path work, and we now have only fifty months within which to exercise our option to convert all Earthian industrial productivity from killingry to livingry products and service systems.

I hoped that *Critical Path* made it clear that lacking the accomplishment of the design-science revolution, while also undergoing the transition into a one-world amalgamation of humanity which we are now experiencing, humans would have been catalyzed only into a world-around social revolution of the same bloody historic pattern of revengeful pulling down of the advantaged few by the disadvantaged many.

I hoped that *Critical Path* made it clear that the accomplished design-revolution's prototypes and developmental concepts now make possible for the first time in history a *bloodless* social revolution successfully elevating all humanity to a sustainable higher living-standard than ever heretofore enjoyed by anyone.

I hope that *Critical Path* made it clear that despite the reality of humanity's option to make it for all humanity, my own conclusion as to whether humanity will do so within the critical time and environmental development limits is that it will remain cosmically undecided up to the last second of the option's effective actuation, knowing that beyond that imminent moment lies only the swift extinction of humans on planet Earth.

The critical path I committed myself to in 1927 was and as yet is that of applying all technology and science directly to accomplishing the mass-produced components for advanced livingry for all humanity, instead of continuing to invest the advanced science and technology inadvertently falling out of the weaponry industry into the livingry tools industries. This critical path was inherently a fifty-year path.

• • •

I will now summarize the last few pages quickly.

The social revolution potential now can for the first time in history realize economic success for all and a comprehensive world enjoyment that involves not revengefully toppling the economically successful minority but elevating all humanity to a sustainable higher level of existing and interacting than any humans have heretofore either experienced or dreamed of.

The now-potentially-to-be-omni-successful social revolution could never before in history have been realized. Until 1970 there had always been enough physical resources but not enough metaphysical resources (of experience-won know-how) on our planet to render the physical technology capable of taking care of everyone at a sustainable, eminently successful level of physical well-being—bloodlessly accomplished and sustainable without the coexistence of either a human slave or working class. Until 1970 it had realistically to be either you *or* me, not enough for both. Since 1970 it has become realistically you *and* me—all else is automated acceleration to human-race extinction on planet Earth.

•   •   •

As we know, when the on-foot soldiers at Crécy stuck their pikes into the ground, points slantingly upward and forward, to impale the bellies of the advancing charges of cavalry, it was social revolution, brought about by design-science revolution. Thus armed with their newly-science-designed pikes, the long-overwhelmed many on foot began to gain emancipation from being overwhelmed by the horse-mounted few. Design did it.

Both the word *revolution* and the words *social revolution* have many meanings, from that referring to the mechanical revolutions of a wheel to the social changing of life-styles which occurred when the horse-mounted and -carriaged few, with their many on-foot servants, stable boys, grooms, coachmen, and the vast slums they drove through were almost entirely design-science-obsoleted by the automobile—mounted many covering vastly more miles and having only a diminishing self-servant-class functioning as some of the riders became the auto-production workers, gas-station attendants, etc.

Greater justice and economic improvement for the many is not always the result of social revolution. The Europeans' guns overwhelming of the American Indian bow-and-arrow weapons was in most ways a retrogressive social revolution implemented by design-science revolution. It is always the design revolution that tips the social scales one way or the other. However, sum totally the combined design and social revolutions ultimately favor the many. Between 1900 and today, 60 percent of humans in the U.S.A. have attained a standard of living far in advance of those of the greatest potentates of 1900 while concurrently doubling the life-span of that fortunate 60 percent.

Never before in all history have the inequities and the momentums of unthinking money-power been more glaringly evident to so vastly large a number of now literate, competent, and constructively thinking all-around-the-world humans. There's a soon-to-occur critical-mass moment when the intuition of the responsibly inspired majority of humanity, in contradistinction to the angered Luddites and avenging Robin Hoods, faced with comprehensive functional discontinuity of nationally contained techno-economic system, will call for and accomplish a world-around reorientation of our planetary affairs. At this critical moment will occur a realization by the responsibly inspired majority that the adequate capacity of the invisible technology to sustainingly support all humanity depends on all the resources, physical and metaphysical, being always and only employed for all of world-around humanity as a completely integrated techno-economic system operating entirely on its daily income principally of Sun-emanating energy. The integrated world-techno-economic system purpose is in contradistinction to a union of 150 autonomously operating nation-states, as with the United Nations. All this can now be comprehensively commonwealth-accounted in time-energy work units. All this can provide regenerative-initiative accommodating access of human individuals to the ever multiplying commonwealth-techno-economic facilities. The degree of individual-initiative computerized access to the commonwealth facilities will be predicated on the demonstrated performance and sustained integrity of the individual's ever-forwardly-anticipatory designing competence.

I have been a deliberate half-century-fused inciter of a cool-headed, natural, gestation-rate-paced revolution, armed with physically demonstrable livingry levers with which altogether to elevate all humanity to realization of an inherently sustainable, satisfactory-to-all, ever higher standard of living.

Critical threshold-crossing of the inevitable revolution is already underway. The question is: Can it be successfully accomplished before the only-instinctively-operating fear and ignorance preclude success, by one individual, authorized or unauthorized, pushing the first button of chain-reacting all-buttons-pushing, atomic, race-irradiated suicide?

The only happily promising recourse of each human individual is to our highest intellectual facilties and their mutual, ego-deflated, unselfishly loving preoccupation with comprehensivity and our employment of the most powerful tools of all:

(A) the family of generalized scientific principles governing the operational design of eternally regenerative Universe itself;

(B) comprehending and effectively employing synergetics, with the books *Synergetics* and *Synergetics II* presenting the comprehensive omni-image-able mathematical coordinate system employed by nature, thus avoiding the mentally debilitating, vast-majority-of-humanity-excluding quasimathematical coordinate system employed by present-day science;

(C) comprehending the major objectives and operating strategies of the major opposing power structures of world politics, their present status quo and probable future trending;

(D) comprehending the fundamentals of economics, of wealth vs. money, of the principal features and functioning of industry, banking, and securities;

(E) comprehending the educational system in general as well as the discovery of the shortcomings of science, engineering, and education in general;

(F) synergetically comprehending "what it is all about," as propounded in *Critical Path* and this book, *Grunch of Giants,* and discovering what our options are to confront imminent race disaster; and

(G) the individual discovery of God by a vast majority of human individuals—not the discovery of religions, but the discovery that each and every individual has an always-instantly-open, no-intermediary-switchboard-authority-to-contend-with, no-interference-of-any-kind, direct "hot-line to God": i.e., the weightless, nonphysical communication occurring teleologically between the differentially limited, weightless, nonphysical, temporal, special-case mind of the individual human and the comprehensively integrated, macro-micro unlimited, weightless, eternal, generalized mind of God.

Index

Advertising, 51, 75
Airplane, x, xvi, xvii, xviii, xix, xxi, xxiv–xxv, 31, 32
Alexander the Great, 65
Aluminum, 31, 79
Apollo Project, xiii
Artifacts, x, xi, xv, 4–5, 12, 86
Atomic bomb, ix, xxi, 3, 39, 41–42, 45, 46, 48–50, 53–54, 83, 90
Atomic energy, 5, 44–45
Automobile, xviii, 31, 43, 55, 89
Balance-of-trade accounting, xxi, 29–30
Bankruptcy, 27, 33, 47–48, 52, 54, 58, 74
Banks: Banking, 25, 26, 73, 91
  failures of, 46–48, 53
  Federal Reserve, 40, 69
  industrialization and, 43, 46–48, 53
  savings, 63, 67, 69, 74–75
Behind-the-scenes power structure. See: Grunch of Giants; Human power structures
Bell, Alexander Graham, 42
Birth rates, ix
Bonds, interest-bearing, 43, 46, 59–60, 66, 67, 68, 73, 74
Brain & mind, 9, 12–13, 84, 85
British Empire, 29, 48
Building industry, xi, xv, xvi, xvii, xi, 42
  See also: Housing industry
Capitalism
  banking and, 43, 46–48, 57–59, 65–76

early forms of, 25, 26–27, 30
  lawyer, 36–37, 50, 82
  post-World-War-II, 2, 38–42, 48, 52–54
  risktaking in, 26, 27, 36, 57–59, 60, 74, 82
  shift to "know-how," 78
  See also: Banks; Corporations; Landownership
Carter, Jimmy, 38
CIA, 49, 53
Communication, xix, xxi, 11, 85
Communism vs capitalism, 3, 82
Computer, ix, xix, 5, 32, 33, 41, 80, 81, 90
Copper, 25, 42, 46
Corporations
  as nonphysical, 18, 28, 29, 34, 37–38, 39–41, 57, 78, 81, 82
  economic games of, 57–60, 63, 67–76, 77–78, 81, 82
  granted privileges of, 29
  industrial games of, 55–57
  mergers & acquisitions of, 33, 37–38
  propaganda of, 57
  royal sanction of, 26
  supranational, xxi, 1–2, 17, 34, 35–36, 40–41, 44–45, 52–54, 80–81, 82
  See also: Capitalism; Grunch of Giants; Profit; Stockholder; Stocks
Critical Path, xxiii, 48, 68, 77, 80, 85, 86, 87, 91

Critical path, 86, 87
Critical proximity, 15
Cross-breeding, 11, 39
Darwin, Charles, 7
David vs Goliath story, 17, 19, 22, 82
Design (of humans), xxiii, 13, 84
Design (of planet), 84
Design (of Universe), 8, 9, 13, 84, 90
Design science (revolution), 15–16, 80, 83, 85, 86, 87, 88–90
Direct experience, 8, 9, 13, 65, 88
See also: Experiment
DNA-RNA, xxiii, 7, 11, 84
Drake, Sir Francis, 28
Dwelling machine, xii, xv–xvii, 79
Dymaxion Bathroom, xii
Dymaxion Car, xiii, 64–65
Dymaxion Deployment Unit, 79
Dymaxion House, xii
Dymaxion Sky-Ocean Map, xii
Earth, as sphere, 27, 84
Eastbound man: Westbound man, 21, 22, 24, 27
Ecology, xxvi, 7–8, 84
Edison, Thomas Alva, 42, 43
Education, 21
Einstein's equation: $E = Mc^2$, xxii, xxvi
Eisenhower, Dwight D., 40–41
Electric utilities, 30, 31, 43–44, 45, 46, 47, 80
Emergence through emergency, xv
Energy accounting, xi, 41–42, 44–45, 46, 89–90
See also: Petroleum
Environment: Altering the environment, 7, 8–9, 11–12, 14, 16, 84–90
Ephemeralization, x, xix, 5, 31–32, 39, 43, 44, 67, 78–79, 80, 86
See also: Performance per weight
Evolution, 13
cosmic, xxiii, 76, 84–85
class-one vs class-two, xix, xxiii, xxiv, 86–88
critical moment in, xxi
Experiment (undertaken by individual), xxiv, 81
Farms: Farming, 20, 47
Food, xxi, 2, 3, 78
Ford, Henry, 55–56
4D monograph, xv
Fourteenth Amendment, 28–29
Frequency in pure principle, 11, 15
Fuller, R. Buckminster
at birth (1895), 66
marriage (1917), 63
U.S. Navy (1917), 63
crisis of 1927, 8
commitment of 1927, 86, 87
4D monograph, xv
Dymaxion House, xii, xvi
Dymaxion Car, xiii, 64, 65
Dymaxion Bathroom, xii, xiv
Nine Chains to the Moon, xiii
Eco-Transformation Charts, xiii
Dymaxion Deployment Unit, 79
Phelps Dodge Company, xiv
on Fortune staff, xii, xvi
Fuller House Company, xvii
Dymaxion Sky-Ocean Map, xii
book shelves, xiv
Octet truss, xiv
geodesic domes, xiii, 32, 64
rowing needle, xiv
World Resources Inventory, 86

World Game, xii

Tensegrity projects, xi

Synergetics, xii, xiv, 91

Philadelphia archives of, xi

annual costs of, 64

See also: *Critical Path;* Self-discipline

"General Houses," xvi

Generalized principles, xxiii, 9, 12, 13, 81, 83, 84, 85, 90

Geodesic dome, xiii, 32, 64

Gestation rate, xv, xvi, xvii, 51, 85, 86, 87, 90

Giamatti, A. Bartlett, 51

Giants, 14, 17, 18–26, 28, 81–82

God, 20, 21, 91

Gold, 25, 29, 36, 69, 70

Good & bad, xx

Gravity, xxiii, 13, 84

Grunch of Giants, 1, 2, 3, 17, 18, 29, 33, 36–39, 40–41, 48, 51–52, 57, 80–81, 82, 91

See also: Human power structures

Heyerdahl, Thor, 21

Horse, 7

Horsemen, 19–21, 22–23, 88

Housing industry, xv, xvi, xvii, xviii, 78

Human individual, xxii

God and, 91

initiative & drives of, xxiii, 10, 85, 86, 90

integrity of, 11

mobility of, 78

success of, xx, 18–19, 80, 81, 85–86, 87–88, 90

vs great nations & corporations, 18

Human power structures

and military supremacy, 41–42, 44–46, 52–54, 91

networking and, xxiii, 89

origin of, 19–23

political-corporate authority as, 52, 57–59, 80–81, 82, 83, 89, 91

selfishness of, xxii–xxiii, 79

See also: Grunch of Giants

Human thought, xxii, 84

Human tolerance limits, 13, 84–85, 87

Humans

as decision-makers, 50

born helpless, 10

extinction of, ix, 45, 75, 87, 88, 91

function of, as information-gatherers & problem-solvers, x, xxiv–xxvi, 9, 10, 11, 85

in Universe, 8, 12, 84–85

integration of, x, xix, xxii, 50, 75–76, 86, 89

selfish, x, 79, 90

tool complexes of, 15, 62–63

See also: Design (of humans)

Humanity in crisis, x, xxviii, 42, 45, 75–76, 87, 90, 91

Humanity's option to "make it," ix, x, xxvii, 3, 5, 6, 41, 76, 80, 83, 86–88, 90, 91

Inbreeding, 7, 13

Industrialization, 2, 37–38, 55–56

Information gathering, xxiv–xxvi, xxvii, 16

Insurance companies, 60, 63, 64, 65–67, 70, 71

Interest, 25, 63, 64, 74–75

Invention. *See:* Artifacts

Invisible: Invisibility

in metallurgy, chemistry, electronics, xvii, 1, 35, 40–41, 83

Invisible (*cont.*)
  of corporate conglomerates, 17, 18, 35–36, 38, 39, 40–41, 51–52, 57, 80, 82
  of technological gains, xx, xxviii, 5, 86
  of utilities, 44–45
  warfare, 39
  See also: Ephemeralization; Scientific & technical know-how
Killingry, 5, 39–42, 51–54, 60–61, 86
Land & sea transport, xiii, xix, 21, 22
Landownership, xxviii, 17, 19, 20, 21, 23, 27, 35, 48, 81
Laser, 5
Law (civil & social), 16–17
Learning through trial-and-error, xi, xxiv, 3, 10, 11, 65, 85
Length-to-surface ratio, xxii, 14
  See also: Volume-weight ratio
Life expectancy, 62–63, 65, 66, 89
Life support inadequacy, xx, xxvii, 3, 6, 66, 82, 87
Light, speed of, xxii, 53, 78
Limited liability (Ltd.), 27, 28, 30
  See also: Bankruptcy, Corporations
Literacy: Illiteracy, xxi, 13, 82, 83, 89
Little individual, xi, 17, 19, 22, 82, 86, 87, 89
Livingry, xi, xv, xvi–xvii, 54, 64, 78, 86, 87, 88, 90
  See also: Service industry
Livingry vs killingry, ix, xi, xviii, 3, 15–16, 44–46, 80, 83, 86, 87
Luddites, 89
MacCready, Paul, 5, 83

Making money with money, xvi, 18, 26, 42–43, 46–48, 66, 82
Mass interattraction, 13, 15
Mass production, x, 24, 30–31, 80, 86, 87
Mathematics, 9, 16, 81, 84, 85, 91
Medicine, 61–62
"Merchant Venturer's Society," 27
Metals, xii, xiii, 25, 78
Milos, 24
Mind, 9, 12, 13, 15, 84, 85, 90, 91
  See also: Brain & mind; Scientific & technical know-how
Mind vs muscle, 8, 17
Miniaturization, 32, 33
Money, 23, 25, 77, 91
Moneymaking, xiv–xv, xxi, 39–40, 55–56, 64, 82
"Moral Majority," 50–51
Morgan, J. P., 42–43, 46–48
Mortgages, xvii–xviii
Nation: National Sovereignty, x, xviii–xxi, xxiii, 76, 78, 86, 89, 90
Navigation, xx, xxiv
Networks, xx–xxiii, 83, 85, 90
Newton's second law of motion, 13
Nine Chains to the Moon, xiii
Nixon, Richard, 52, 69, 73
One world family, x, xviii, xix, xxi, 76, 86, 87, 89
Operating Manual for Spaceship Earth, 26, 77
Performance per weight, x, xiii, 4–5, 31–32, 43–44, 49, 79, 86
Petroleum, x, xx, 5, 43, 45
  See also: Energy Accounting
Physical & metaphysical, 10, 12–13, 16, 17, 18, 78, 91

Pirates, 21, 27, 29
Politics, ix, x
    See also: Revolution
Population growth, 3
Prime contractors, 33, 52–54
Problem-solving, x, xii, xxiv–
    xxvi, xxvii, 3
    See also: Humans
Profit, 30, 68
    corporate, 28, 42, 55
    for all, ix
    from ship ventures, 26–28
    motive, xxvii, 39–40, 42, 55–56
Prototyping, x, xv, 86, 87
Pure principle, 15
Pygmies, 14
Queen Elizabeth, 28
Race: No race, no class, x
Radiation, xxiii, 84
Radio-triangulation mapping, 42
Railroad industry, xvi, 30, 43
Reagan, Ronald, 37, 40, 50,
    73
Reforming the environment not
    humans, xi, xx, 63, 86
Regeneration, xx, xxiv, xxv, 7–8,
    15, 16, 90
Revolution (political), 83, 85, 86,
    87, 88
Revolution (social), 85, 86–88,
    89, 90
Roland (Childe Roland), 19,
    23
Roosevelt, Franklin Delano, 48
Sailing ships, xiv, xix, 23
Satellite, xix, 5, 83
Scientific & technical know-how,
    xx, xvii, 2, 5, 78, 83, 87
Securities. See: Bonds
Securities & Exchange Commis-
    sion (SEC), 59
Self-discipline, 8

of artifact development, xiv
of avoiding moneymaking as
    goal, xiv
of being apolitical, x, xxiii
of learning from mistakes, xi, x,
    10, 11
of reforming the environment,
    not behavior, xi, xx, 86
of socioeconomic interests, xix
Service industry, ix, xvi, xvii,
    86
Ship, 24, 26–27, 31–32
Ship technology, 23–24
Silver, 25
Socialism vs free enterprise, 38,
    54
Socializability of resources, 42
Solid: No solids, 15, 81
Spaceship Earth, xx, xxi, 80
Specialization: Overspecializa-
    tion, xxviii
Specialized equipment, 8–9, 14–
    15
Standard of living, higher, ix, xi,
    xx, xxvii, 5, 54, 80, 86, 87–
    88, 89, 90
Steel, 46, 57
Stockholders, 27, 28, 29, 30, 35,
    37, 40, 44, 55–58, 66–67, 68,
    70, 79
Stocks (investments), 46, 57, 66–
    69, 70, 72, 73–74, 84
Submarine, xiii, 49, 61
Sun energy, daily income of, xi
Survival, x, 2, 3, 5, 6, 7–8, 15, 41–
    42, 44–46
    See also: Humanity
Synergetics, xii, xiv, xx, 15, 81,
    84–85, 91
Tax, 67, 70, 72–75
Technology, ix, xiii, xviii, xx, 1,
    16, 17, 30, 40, 44–46, 80

Technological revolution, xx, 80, 83, 86
  See also: Design science
Telephone, xviii, xix, xxi, 42, 63, 80
Tensegrity, xi, xiv
Thatcher, Margaret, 40
Things: No things, 15, 81
Tin, xiv
Tool, 9, 11, 14–15, 16, 17, 24, 31, 58, 87, 90–91
Touching: Never touching, xi, 13, 15
Trade, 23, 25, 27–28
Trafalgar, Battle of, 29
Trust funds, 43, 67–69
Truth, 10, 11, 81
Turner, Stanisfield, 39
Turpin, 19
United Nations, 29, 36, 90
United States of America
  defense strategy of, 39–41, 44–46, 48–50, 51–53
  economic history of, 33–54, 61–62
  elections of, xxiii, 45, 50
  national debt of, 35, 44–45, 52–53, 74, 76
  population of, 50
  prestige declines, xxiii, 35–36, 68, 74
  Social Security of, 70–71, 73
  trustbusting of, 37–38, 57
U.S. Department of Defense, 34, 49
U.S. Justice Department, 37

U.S. Supreme Court, 28, 29, 45
U.S. Treasury Department, 70, 73–75
Universe, xi, xxvi, 3, 9, 12, 13, 16, 84, 90
  See also: Design
Union of Soviet Socialist Republics, 48–50, 53, 54
U.S.S.R.-U.S.A., 3, 35, 41–42, 45–46, 48–50, 53–54
  See also: Killingry; Livingry vs Killingry; World War
Up & down, 81
Volume-weight ratio, 14
War, as obsolete, x, 5
Water-people, 21
Wave behavior, xviii–xix, xx
Wealth, 25, 26, 30, 43, 44–45, 77, 91
Weaponry, 15, 53–54, 60–61, 88–89
Weaponry industry, xxi, 15, 31, 33, 39–42, 44–46, 48–50, 52–54, 72–73, 87
  See also: Killingry
Weinberger, Caspar, 49
Word, as tool, 11
Working class, 10, 17, 24–25, 88
World Game, xii
World War I, 34–35, 48, 52, 61
World War II, 35, 38, 44–45, 53–54, 62
World War III ("cold") and IV (total), 35, 38, 39, 41–42, 44–46, 48–50, 53–54, 90
  See also: Atomic bomb